George Frederick Pentecost, Edward Hooker Dewey

The True Science of Living

The New Gospel of Health - Practical and Physiological

George Frederick Pentecost, Edward Hooker Dewey

The True Science of Living
The New Gospel of Health - Practical and Physiological

ISBN/EAN: 9783744669573

Printed in Europe, USA, Canada, Australia, Japan

Cover: Foto ©berggeist007 / pixelio.de

More available books at **www.hansebooks.com**

THE TRUE SCIENCE OF LIVING.

THE NEW GOSPEL OF HEALTH.

PRACTICAL AND PHYSIOLOGICAL.

STORY OF AN EVOLUTION OF NATURAL LAW IN THE CURE OF DISEASE.

FOR PHYSICIANS AND LAYMEN.

HOW THE SICK GET WELL; HOW THE WELL GET SICK.

ALCOHOLICS FRESHLY CONSIDERED.

"Health is the First Wealth."—*Emerson.*

"The physician who wants to know man, must look upon him as a *whole*, and not as a piece of patched up work. If he finds a part of the human body diseased, he must look for the *cause* which produced the disease, and not merely the external effects."—*Paracelsus.*

BY

EDWARD HOOKER DEWEY, M. D.

INTRODUCTION BY

REV. GEORGE F. PENTECOST, D. D.

NORWICH, CONN.
THE HENRY BILL PUBLISHING COMPANY.
J. & J. BUMPUS, LIMITED, NO. 350 OXFORD ST.,
LONDON, ENGLAND.

1895.

INTRODUCTION.

THE beloved apostle, writing to his well-beloved Gaius (3 John ii.), whom he loves in the truth, greets him thus:

"Beloved, I wish above all things that thou mayest prosper and be in health, even as thy soul prospereth."

Spiritual prosperity and health, which John takes for granted in his salutation to his beloved Gaius, is certainly the highest blessing attainable in this earth; but John prays that this great blessing may be matched by another, namely, the prosperity and health of the body. It is with the sincerest desire that the readers of this book may improve their health and increase their prosperity in body, soul, and spirit that I have most willingly set my hand to write a brief introduction to the pages of Dr. Dewey's book.

Bodily health is certainly desired by all men and women, especially by those who have suffered from any loss of health or impairment of physical strength. For the most part bodily health is desired as a principal factor in our earthly enjoyment, and for the sake of earthly gain and prosperity. But the Christian ought to desire health of body for the higher reason that so he can serve God the more efficiently. The body is the Lord's as well as the soul and spirit. " He is the

3

Saviour of the body also." "Know ye not that your body is the temple of the Holy Ghost?" To defile the body with sin, or to voluntarily neglect the body in anywise as to cause it to suffer in health or strength, is an offence against our salvation and the honor of God. To deliberately undermine the health and strength of the body by persisting in an injurious, because false, way of living, is a sin of great enormity. Drunkenness or gluttony are offences against both body and soul which no self-respecting person, not to say Christian, ought for a moment to allow. Nevertheless, there are no doubt hundreds and thousands of good men and women who are guilty of both excess and bad methods in their eating and drinking that border upon these two disgusting and harmful sins.

The object of this book, as I understand it, is to put before the reader a "better way of living" than that which characterizes the great majority of people. It relates to the habit of eating and drinking, and sets forth from the point of view of sound physiology the relation of food to the human system, and so to disease and health.

If the author of this book is right in his premise and conclusion, he has set before the world a theory of living which ought to, and I believe will, revolutionize the habits of a multitude of right-thinking men and women. If he is right in supposing the coursing through our veins of pure, rich blood goes not only to preserve the health of the body and prolong life, but also that it makes for righteousness, both in clearing the mind and purifying the body of those humors which make fuel for unholy desires and unrighteous dispositions, then we must give attention to what he says. "Pure and rich blood (the life is in the blood) contributes to the moral, intellectual, and spiritual growth as well as to the bodily health." This proposition is almost self-evident and needs no argument. We should hail, then, with gladness, any discovery that will enable us to purify and enrich our blood.

Of the two principal matters recommended as the practical outcome of the theory of health developed in this book, the first is that fasting, or the abstinence from food until natural hunger calls for it, is the best way to bring about recovery from disease. Take away food from a sick man's stomach and you have begun, not to starve the sick man, but the disease. We have all of us heard that fevers are more difficult to subdue in large and full-habited people than in the "lean kind." I have often had my physician tell me that if ever typhoid or pneumonia got hold of me it would "go hard with me," for the reason that there was so much material for the disease to feed upon. A conflagration is great or small in proportion to the quantity of easily-inflammable fuel the fire has to feed upon.

The second is that digestion is best promoted and food so assimilated as to afford the largest amount of nourishment and the greatest quantity of rich blood, by giving the stomach a long rest from all work during each twenty-four hours. That is to say that we shall all be the better by giving the stomach rest from the evening till the noon of the next day. In other words, Dr. Dewey recommends that we should give up our breakfasts, and by so doing we are certain to improve our health if we are well; prevent the incoming of disease; assist nature to recover from any unavoidable attack of sickness; strengthen the whole body, and thus build up the soul and the spirit which are so intimately connected with the body.

The only right I have to commend the book is the right which one man has who has tried the "better way of living" and has found that "it works well."

I am not personally acquainted with Dr. Dewey, the author of this book, but I have had the pleasure of some correspondence with him, and have experienced the benefit of greatly improved health by following out the simple rule of "right living" which he lays down in his pages. I am therefore quite ready to put my hand to an unconventional introduction, if by so doing I shall be enabled to induce any one to read these pages carefully, and, so far forth as the things therein contained applies to him, to follow the advice given faithfully.

When the blind man was asked by the Pharisees how Jesus opened his eyes, he related the fact of the process by which the miracle was wrought without attempting to explain the mystery of power underlying it. He answered and said, that "A man that is called Jesus made clay and anointed mine eyes and said unto me, Go to the pool of Siloam and wash; and I went and washed, and I received sight."

And again when the Pharisees cross-examined him he repeated his simple testimony,

"He put clay upon mine eyes and I washed and I do see."

I am but a layman in the medical science and so do not pretend to discuss the subject professionally, but, as "a grateful patient," I am desirous of testifying to the benefits I have received from following Dr. Dewey's method of "right living."

This I know, that for forty years I have been a miserable victim of sick headache, induced by "a kind of indigestion," by "a torpid liver," by this and that, as I have been told by many physicians. I have tried every remedy and expedient that has in turn been recommended to me by physicians and friends. In many of them I have found temporary relief; but the cause of the trouble has ever remained, and the bilious sick headache, with its excruciating pain, would return and a total collapse of my power to work would supervene for from one to three or four days. I have tried dieting, that is not eating so heartily, not eating certain kinds of foods, not drinking coffee, etc. I have tried exercise of various kinds. I have tried preventive remedies, in the form of sodas of various kinds, antipyrins, antifebrins, blue pills, bromides of various kinds, etc. I have tried Turkish baths, and massage. All these things have given me more or less temporary relief, but I have always known that it was but temporary; that the real trouble was untouched. In addition to this bilious habit with its dread accompaniment of headaches, I have been steadily gaining in weight for twenty years past, until I had reached the great weight, for a man of my height (5 feet, 9 inches) of two hundred and fifty pounds. This has of course inconvenienced me, and brought on a certain shortness of breath upon the most moderate exertion either in walking or running, especially in running and going upstairs. I would

not like to give the idea to any reader that I have
been in any wise a sick man, for I have never, with
the exception of the times when for a day or two I
have been laid aside with sick headaches, been in bed
a week in my life of fifty years, with any kind of
sickness or disease according to the ordinary accepta-
tion of these words. Indeed I have been all my
life a man of extraordinary health and strength, doing
tremendous work in the line of my calling, preaching
daily for months or years together to great crowds
of people in every part of the world. At the same
time I have always been conscious of the fact that
there was serious trouble behind this great store of
health and strength, and especially has the steady
accumulation of fat in my system been a source of
anxiety as well as discomfort to me. The tendency to
vertigo and a flushed face, and at times great lassi-
tude which I could only overcome by great effort of
will, has also caused me anxiety. I have been warned
more than once by my doctors that I ought to be very
careful not to make any great or violent exertion, as I
was liable to suffer at any time from suffusion of
blood upon the brain.

Well, some months ago, I chanced through a friend,
whom I had known to be an invalid for years, and
whom I then saw in seeming perfect health, to hear
of Dr. Dewey and his method of "right living." I
found that not only my friend, but every member of
his family, including an invalid wife, a delicate
daughter, two splendid young collegians, and a young
boy of twelve had all given up eating their breakfasts ; and that they were all greatly improved in

health and strengthened mentally as well as physically. I was introduced by my friends to several other persons in his city who had adopted the "right-living" method, and with one accord they all testified to the same great benefits experienced. I called on one or two business-men of my acquaintance who had adopted this method of living, and being men of my own type, they testified that they had, one and all, lost their tormenting sick headaches, lost a great deal of superficial fat and tissue, and were in every way greatly improved in health, in spirits, and in their capacity for work. I called upon an eminent physician whom I had known and who had on one or two occasions prescribed for me. I asked him if he knew of Dr. Dewey's method of treatment and living. He said he did, and strongly recommended me to a adopt the anti-breakfast régime and confessed, *sub rosa*, that he himself had adopted it, and was greatly the better for it. I learned of friends in my own calling who had suffered for years on Monday with fearful headaches as a result of physical and nervous exhaustion incident on their Sunday's work, who, having given up their breakfasts, had recovered entirely from the dreadful Monday prostrations and were enabled to do more and better work than ever before both in their studies and in their pulpits.

Taking the theory upon which this system of living is based into account (and even to my lay mind it seemed most reasonable), and the testimony which I personally received from both men and women, delicate and biliously strong, working-men, merchants, doctors and preachers, delicate ladies for years in-

valided and in a state of collapse, and some who had
never been ill, but who were "an hundred per cent.
better" for living without breakfast, *I resolved to
give up my breakfast.* I pleaded at first that it might
be my lunch instead, for I have all my life enjoyed
my breakfast more than any other meal. But no! it
was the breakfast that must go. So on a certain fine
Monday morning I bade farewell to the breakfast-
room. For a day or two I suffered slight headaches
from what seemed to me was the want of food; but I
soon found that they were just *the dying pains of a
bad habit.* After a week had passed I never thought
of wanting breakfast; and though I was often present
in the breakfast-rooms of friends with whom I was
visiting, and every tempting luxury of the breakfast
was spread before me, I did not desire food at all,
feeling no suggestion of hunger. Indeed now, after a
few months the thought of breakfast never occurs to
me. I am ready for my lunch (or breakfast if you
please) at one o'clock, but am never hungry before
that hour.

As for the results of this method of living I can
only relate them as I have personally experienced
them.

1. I have not had the first suggestion of a sick
headache since I gave up my breakfast. From my
earliest boyhood I do not remember ever having gone
a whole month without being down with one of these
attacks, and for thirty years, during the most active
part of my life, I have suffered with them oftentimes,
more or less every day for a month or six weeks at a
time, and hardly ever a whole fortnight passed without

an acute attack that has sent me to bed or at least left me to drag through the day with intense bodily suffering and mental discouragement.

2. I have gradually lost a large portion of my surplus fat, my weight having gone down some twenty pounds, and my size being reduced by several inches at the point where corpulency was the most prominent; and I am still losing weight and decreasing in size. The process of reduction is very gradual, but still is maintained from week to week.

3. I find that my skin is improving in texture, becoming softer, finer and more closely knit than heretofore. My complexion and eyes have cleared, and all fullness of the face and the tendency to flushness in the head has disappeared.

4. I experience no fullness and unpleasantness after eating as I so often did before. As a matter of fact, though I enjoy my meals (and I eat anything my appetite and taste call for) as never before, eating with zest, I do not think I eat as much as I used to do; but I am conscious of better digestion; my food does not lie so long in my stomach, and that useful organ seems to have gone out of the gas-producing business.

5. I am conscious of a lighter step and a more elastic spring in all my limbs. I can walk with quickness and for longer distances without consciousness of "that tired feeling" I used to experience. Indeed a brisk walk now is a pleasure which I seek to gratify, whereas before the prescribed walk for the sake of exercise was a horrible bore to me.

6. I go to my study and to my pulpit on an empty

stomach without any sense of loss of strength mentally or physically,—on the other hand with freshness and vigor which is delightful. In this respect I am quite sure that I am in every way advantaged.

I may add, that, after seeing the manifest improvement that had taken place in my whole physical condition, my eldest daughter determined to follow in my footsteps, and she even went so far as to suggest to her little son (seven years old), that he also give up his breakfast. At first the wee chap said he wanted his breakfast and had a cry at the thought of being deprived of one of his natural and habitual rights. But his mother explained to him that she thought it would be a benefit to his health, and gave him a few simple reasons for the advice she had given him, and the little fellow pleasantly gave in to his mother. Nothing will induce him to eat his breakfast now till one o'clock, and both he and his mother are much better in health than before. When I returned to England some months ago from America, my son, a lad of twenty, asked me why I did not eat my breakfast and I explained to him as best I could the theory of the " better-living " system, beginning with the proposition that " restful sleep is not a hunger-causing process," and expounded also the advantage of a long rest for the stomach. I also pointed out to him that it is not the *quantity of food* which one eats that produces blood (good blood) and strength but the *amount of nourishment* which we get out of the food. After an hour's talk over the matter, without any recommendation or even suggestion on my part, and much less without thought that he would adopt this better method of living (for he was a great devotee to his breakfast) he said of his own accord:

"Father, that seems a most sensible theory of living. I shall give up my breakfast at once."

Since then he has eaten no breakfasts, and testifies to the fact that he feels a good fifty per cent. better all the day through, and does his (office) work much easier and with a clearer head than before. He now playfully speaks of breakfast to his friends as "that vulgar habit."

One by one, of their own accord, every member of my family have given up their breakfast, and I think I can safely say that all are the better for it, though with one or two of them the sacrifice of the breakfast-god has been of so recent a date that I can only say they are the better on the principle of inductive philosophy.

My friends have, almost without an exception, noted and remarked upon the great improvement in my general health and appearance; and almost invariably said something like this:

"How well you are looking." "Your holiday has done you a world of good." "Why, what have you been doing with yourself this summer? You are looking better than I ever saw you." "You must have been visiting the fountain of youth this summer. Tell me where it is and I will make a pilgrimage!"

To these and similar remarks I have simply replied: "Oh, I have discovered the 'better way of living' and been following it; and to that alone I attribute my general improvement in health and appearance."

"What is it? Do tell me, for really I am so dull

and tired that life is sometimes a burden to me."
(This remark is literally quoted.)

"Well, it is very simple.   Just give up eating your
breakfast."

"What, give up my breakfast!"   "Why! I would
rather give up any other meal in the day."   "Besides
I could never do my work without my breakfast."
"I should faint before ten o'clock."   "It is quite true
that I eat very little breakfast, but I am sure I could
never get through the morning without it."   "Oh, that
may do very well for a great healthy man like you, but
I could never manage it."

These are specimen answers, quoted literally; to
which I have replied invariably by a brief statement of
the anti-breakfast theory and then expounded; begin-
ing with "Restful sleep is not a hunger-causing process,"
and going on to suggest that the breakfast they eat
cannot furnish strength for them to do their morning
work upon, as in no case does food give strength until
it is converted into blood.   That we all do our work
to-day largely if not entirely on the blood extracted
from the food of yesterday.   Then I have questioned
them as to their habits and state of health, and in nine
cases out of ten I have found that these friends of
mine, the more healthy ones as well as the delicate
ones, all suffer in some degree from some one of the
many forms of indigestion.   Headaches, palpitation,
fullness of habit, neuralgia, accumulation of gas in the
stomach, pimples on the face, or some other form of
eczema arising from and due to poor blood.   These
conversations have always awakened interest, and even
the most skeptical have again returned to the discussion

of the matter, as though the very mention of the matter had produced conviction.

In at least a score of cases my friends have adopted the "right-living" method, and all testify to their great delight in it. In one or two cases where dyspepsia has been a veritable fiend, and health and strength almost gone, with depression of mind and even decay of will, the improvement has been most marked, and these friends are on the high-road to health again. I could fill pages of interesting details coming under my own observation and resulting from the adoption of the rule, on the strength of my simple expositions, which would be surprising.

I am often asked if it is possible that I preach on Sundays, that is in the morning, without any food, as though that were a feat almost incredible. To which I am able not only to say, "Yes, of course," but more than that, "I go to my pulpit fresher in body and mind, and come out of it fresher after the sermon, than I ever did in the old breakfast-eating days."

I have come to the conclusion that the blood cannot take care of the brain and the stomach at the same time, and if a man has a breakfast to digest and a sermon to preach during the same period of time, either the breakfast or the sermon will have to suffer, and most likely both of them do. So convinced am I of this fact that I am almost prepared to believe that preachers would do better work and be stronger in body if, as a rule, *they took no food at all on Sundays,* but only drank what they cared for of water. A twenty-four hours' fast from all food once a week would not only do no harm, but would give the

stomach such a rest as would enable all the other
bodily functions to clear the body of unhealthy
remainders. God has ordained a seventh-day fast from
exhausting and secular labor. Of old He ordained a
fast once in seven years for the land that it might
recover and recoup itself from too constant labor of
production. We know that all ship-owners and other
users of machinery require their engines and boilers to
have an occasional rest. Two pairs of boots will wear
longer, if the use of them is alternated, than three pairs
worn steadily, one pair after the other till each is
worn out. Why then is it unreasonable to suppose
that it is a good thing to give the stomach, that most
delicate and important of all our organs, as long a rest
each day as possible, and occasionally a much longer
rest!

Some of my friends have charged me with having
fallen into the hands of a "quack," and have thrown
this at me:

"If this anti-breakfast theory for people who are
not invalids, and the 'starvation theory' for sick people
were true, do you suppose that it would have been left
for an obscure country doctor in America to have discovered it? Have any of the great medical authorities,
such as Sir A. B. and C. D. and Q. X., recommended
it?" In a word, "Have any of the rulers believed on
him?" I do not pretend to argue the matter techni-
cally as I am not a technically-educated physician.
Therefore my testimony is that of a layman. As such
I give it for what it is worth. I cannot help, however,
thinking again of the case of the blind man who was
questioned by the Pharisees. They had a theory that

Jesus must be a bad man and a sinner because He healed on the Sabbath day, which to them was the most sacred thing they possessed, more sacred than God Himself. The man that was born blind and yet was restored to sight replied very wisely, as a layman:

"Whether He be a sinner or no I know not; one thing I know, that whereas I was blind, now I see."

It is even so in such a case as this. Here is a man with a theory for "better living," which he is prepared to defend on scientific principles, and to demonstrate by actual experimental evidence. Here is a theory of "better living" which I and scores of others have tried, so simple, so full of common sense, and withal one which we have demonstrated by a simple experimental test. The theory may not be indorsed by the medical profession, nor widely acted upon by practitioners, but since "it works well" with all sorts and conditions of patients, we are bound to say again with the blind man, modifying his words a little. "Why, here is a marvelous thing, that ye know not whence this 'better way of living'" comes from, yet it giveth better health and enables nature to cure innumerable diseases that have in other patients proved fatal, simply by letting her take her own course and not worrying her by over-feeding a diseased stomach and lashing it as a cabby does his tired and jaded horse.

For the benefit of my fellow-ministers, into whose hands this book I hope will fall, I pass Dr. Dewey's prescription to me, on to them.

"Always go into your study, your pulpit, and your bed with an empty stomach. Follow this rule as

2

nearly as you can, and I will guarantee the largest measure of health and strength that is possible in your case."

In any case I most seriously and heartily recommend that one and all of the readers of this book *give up eating breakfast,* and they will know in themselves in less than two months whether the doctrine be based on sound principles or whether it be the vagary of a quack. " The proof of the pudding is in the eating." It certainly will do no one any harm to leave off the breakfast for three months and it is equally almost certain that before that time has elapsed any one so doing will need no further argument.

GEORGE F. PENTECOST.

LONDON, ENGLAND,
     November 9, 1894.

CONTENTS.

PART I.—NATURE IN DISEASE.

CONTENTS.

PART I.

NATURE IN DISEASE.

NATURE'S BILL OF FARE FOR THE SICK.

FAT	91 per cent.
MUSCLE	30 per cent.
LIVER	56 per cent.
SPLEEN	63 per cent.
BLOOD	17 per cent.
Nerve Centers	0 ! ! ! ! !

LECTURE I.

INTRODUCTORY.

ARMY LIFE. ELEVEN YEARS OF PRACTICE, INVOLVING A SLOW AP-
PROACH TO NATURE IN DISEASE. DOUBTS AS TO THE EFFICI-
ENCY OF REMEDIES AS GENERALLY USED, WITH RESULTING
ABRIDGMENT OF THE AUTHOR'S MATERIA MEDICA.

My Friends the Readers :—

I have invited a few of you into my little private
lecture-room where I can talk to you at short range,
and where for the purpose of intellectual association
you are to become my listeners. I have invited you in
particular because you have receptive and therefore
responsive minds. You are good listeners, and good
listeners you know are always at the intellectual level
with the speaker whether his plane be high or low,
whether he talks sense or nonsense.

To talk to you as I could wish to impress you, as I
wish to impress you, I must get very near you, so near
that whenever a live thought is so received as to
become instantly vitalized, I can see the glow of
your countenances, the sparkle of your eyes, that I
may realize a reflex stirring up of my own mental
and moral powers, and I want to be so near you that
you can see the strong conviction behind the strong
expression.

I choose the morning as the best time for meeting
you, because the play of the mental faculties is easier

and stronger every way both for me to give out and for you to receive after the night of physical, restful regeneration.

I am going to tell you *how to feed the sick, even the very severely sick,* and even with intense aversion to food, so as to reach the highest possible condition of support to vital power. I am going to tell you how this is done without ever a mistake as to the kind, the quality, or the amount of food to be assimilated. I am going to tell you how this food is made available so that it is drawn upon to support vital power in exact proportion to the need, without the slightest taxing of digestive power.

And in connection with this method of feeding the sick, I shall discuss the use of stimulants as remedies. I shall present to you who are crusaders against alcoholics two new arguments against their use. I shall try to make you who are not crusaders, believe that from the time that Noah planted the vineyard, drank wine and was drunken, down to the present, *every dose of an alcoholic that ever went into a human stomach has had its degree of debilitating effect upon vital power, and by so much has become a hindrance, and not a support in time of disease.* I shall try to make you believe that every dose of an alcoholic that ever went into a human stomach has had its degree of local irritant effect that has proportionately lowered its functional power.

If you admit that alcoholics are bad for the well, because of their effects on the brain and stomach, and worse for the sick, because of a decline in defensive power against these effects, then I shall try to make you believe for the same reasons that their use is *positively dangerous* in all cases of shock from injury, and in every crisis of disease.

I am going to tell you of a remedy to create hunger, a genuine appetizer, that is not to be found in any materia medica, that rarely if ever appears in written prescriptions, and never in conspicuous type in authorized works on the practice of medicine, though its name is known to all peoples. It is a remedy or a means that you will habitually make use of when you become aware of its natural power and effectiveness, because it is available to all, absolutely safe in its operation, and never fails to cause the keenest hunger where death is not inevitable. And by its habitual use eating becomes elevated to a luxury of life, and not only enables you to habitually eat more food on the average than before, but it has the supreme merit of permitting *Nature* to make the bill of fare.

Although the remedy is known to you all, the precise way that I make use of it very nearly amounts to an original discovery. In its general application it involves a physiological plan of living by which your health habits will become automatic, a plan that avoids the necessity of worrying thought as to what you must or must not do for the sake of health, what you must or must not eat in order to have your bodies duly nourished—a plan that will relieve in the highest degree perplexing anxiety about the health, depressing apprehension as to the possibility or the probability of disease—in short, I shall try to make you see that your ordinary ways of living are suicidal to a degree that you are not in the least aware of.

My course of morning lectures will consist, essentially, of a story of an evolution in medicine that began with me, even before I ever saw a medical text-book. Hence what I am to tell you is not what I have found out by delving through medical books in a search for borrowed facts and conceptions to be re-dressed and pre-

sented as my own. I am to tell you what I have
learned through original investigation, not from books
but from Nature herself, while sitting at her very
feet. For science advances only through original investigation.

I begin my story with a slight thread of Autobio-
graphy. I am a son of New England from many generations, but was born in the country among the hills
of Western Pennsylvania where the freshest air is
always available. I grew up in intimate acquaintance with all kinds of manual labor, and was an ardent
lover of Nature as revealed in rocks, hills, woods,
storms, growing crops and in the many-hued, varied
forms of the floral world. Mentally, the loftiest cheer
was the habitual need of my life, and a need so pressing that a sick-room was one of the last places I could
be induced to enter, and conversation about the sick
I always instinctively avoided.

My working tastes strongly inclined me to mechanics,
hence, that kind of farm work that involved the best
application of mechanical force was my preference.
The carpenter's shop was the delight of my boyhood
and youth, and no boy's eyes were ever more apprecia-
tive of the finest exhibition of the miracles wrought
with edged tools. I was the only boy in my entire vicinity who was able to make his own carts and wagons.
My preference in all kinds of labor was strongly for
those where there was the best adaptation of the means
to the end, and where the thing to be done was worth
all the brain and mechanical force expended. Science
means exact adaptation of means to ends.

Before I began the study of medicine I had become
aware of the fact that the majority of persons attacked
by acute disease recovered, no matter how treated or by
whom. In those earlier times the law permitted any one

to practice the dispensation of drugs, roots and herbs without requiring him to know anything about their medical properties, or of the diseases for which he prescribed ; yet the most ignorant of remedy vendors often had success in treating them.　It was often within the range of my knowledge that poor families living at a distance from physicians treated their sick and at times their severely sick with only "home remedies"; and I wondered over the mystery of means, over the mystery of cure.

Now you are all ready to ask, how with such me· chanical tastes and such instinctive aversions it was possible for me to make the practice of medicine my life-work.　"There is a destiny that shapes our ends." It came about in this way.　My father had no extra farms to put his several sons upon, and I was of too slight build to be a success at heavy manual labor. There was a doctor of wise mien and stately form, who was in want of a medical student, and he knew of a drug-store in want of a young man to learn the business, and a knowledge of the drug business would be an excellent preparatory acquisition to the medical department of general practice.　Hence, I could read medical books under his tuition, and become a doctor almost without being aware of it.　An extravagant statement of the incomes of doctors, that in his language were princely compared with those of laborers and mechanics, settled the question of what was to be my business during life.　Through the opened door I walked, never to go out again.

But there was another force behind all that seemed so untoward.　There was a longing, and a strong ambition for a more conspicuous life than could be realized on a farm or in a workshop ; and who more generally conspicuous, more honored, more revered than the trusted

family physician ?   Does he not walk the ways of the
sick-room "every inch a king," with his subjects
ready to render abject obedience to his imperial nod,
whether they be presidents, emperors and kings ; or
whether they be the lowliest plebeians ?

I spent two years behind the drug-counter, during
which I came in contact with all kinds of physicians
and all kinds of "isms" in medical practice.   The
prescription case is a wonderful revelator of the literary
and scientific attainments of the medical profession,
but it fails to account for the relative degree of success
among physicians in a business sense.   It did not
account for the fact that very often men who were
without the shade of scientific conception of the action
of a remedy, or of the indication of its need as revealed
by symptoms, yet were capable of getting a large
business and holding it, and seemingly with the average success of their more learned brethren.

My life in the drug-store did not lessen the mystery
of the power of dosage, in the cure of disease, nor
hinder a slowly developing conviction that, as an
adaptation of means to ends, the administration of
drugs to the cure of disease to a very great degree is
one of the most unscientific of all human avocations.
But I was not hindered thereby from going on with
my studies.

On leaving the prescription case I entered the office
of one of the most learned of the city physicians, to be
further guided in the mysterious ways of medicine.
In due course of time I received my medical degree
from the College of Medicine and Surgery of the
University of Michigan.   I began my professional
career on a bright May morning in 1864, by assuming
charge of a ward in the Field Hospital at Chattanooga,
Tenn., as an Acting Assistant Surgeon, U. S. A., with

its eighty cots, filled with sick and wounded, most of them fresh arrivals from the Battle of Resaca, Ga., that had just been fought, and many of them severely sick and severely wounded.

My previous clinical experience had been very slight. I had never cared for a severely sick person or for a severe wound, yet there were eighty patients thrust upon my inexperienced hands, to encounter possible danger little less than that they had met on the field of battle. They were as helpless in my hands for weal or woe as if I were the only surgeon in the universe, for only ignorance or flagrant violation of duty on my part could release them from my grasp. There were more than a thousand patients in the hospital under the care of a number of ward surgeons, with a surgeon in charge over all ; his duties being executive he knew only in a general way what was going on in the several wards. You may well wonder at the exigencies of the government when a charge so grave could be placed in hands apparently so inadequate ; and also at my temerity in accepting the charge, with such an exceedingly confused state of mind over the question of dosage in the cure of disease. All this seems more impressive to you as you recall the commotion caused by even *one* severe wound or case of sickness in a quiet community, and you feel a trembling sense of sympathy for both the surgeon and his possible victims. I will assure you that those patients were as safe, even on the first day of my care, from any careless or dangerous dosing, as were the patients of the most experienced surgeon in the hospital. My bump of caution had a facility always of developing according to the need ; and though no young man ever felt a keener and more painful sense of self-insufficiency on beginning his professional career than did I on that morning when I began the most trying and taxing day's task in all of my human history, the sheet-anchor that held me to my work, was the one fact that most of my sick would get well anyway, and the wounded would need but little of drugs.

There was one form of treatment always needed, always available, that I could bestow with the ease and effectiveness of an expert even on the first day of my service. I could give kindly and attentive care alike to all as the need was found, and always in full measure.

Here I will point out some of the striking professional advantages of such a service.

1. There were no friends at hand with very natural but very harassing anxieties to be regarded, which in civil practice are often more exhaustive to the vital powers of the physician than the care of the patient.

2. There could be no harassing apprehension as to a discharge from the service in case its success failed to meet the conception of the patient or of the friends, as in civil practice. There was no question of reputation involved as to success or failure, for the service was secure and the remuneration fixed and certain ; hence the surgeon could walk his ward with the easiest possible play of his mental faculties.

3. To the respect naturally incited by professional attainments and to service conscientiously performed, was added that due to superior rank that military life calls out. This also would have its effect on the play of the mental faculties.

4. Postmortems were the rule in this hospital, and as they were more or less numerous every one of the hundred days of smoke and flame of Sherman's advance on Atlanta, the opportunity of finding out how little we knew of the variety, and extent, of diseased structures in our patients while alive, was most extensive. When we had a case of pneumonia, of pleurisy, or of

typhoid fever die on our hands, it was very generally found that it had been involved with diseased conditions, that had given no hint of their existence during life, and had without much doubt rendered death inevitable from the beginning. But there was one fact revealed in every postmortem of tremendous significance that failed to make any impression on my mind other than to remember it. The fact that no matter how emaciated the body, even if the skeleton condition had been reached, the brain, the heart, the lungs, except themselves diseased, *never revealed any loss.* You will keep this fact in mind, as it will be duly considered in another lecture.

5. A striking advantage to the patient as well as to the surgeon was the possibility of the most skilful use of the surgeon's knife with corresponding results. Among the ward surgeons were several operators of great skill and large experience, and, as all the surgeons were expected to be present when severe operations were to be performed, and as no operation would be performed without a general concurrence of opinion, the patient was always likely to have the wise thing done for him.

And another and very striking advantage was the high character of the social relations existing among the medical officers, and there being no cause for professional jealousies, none existed; hence we could ask advice and consult with a degree of freedom not always safe in civil practice.

"A question, doctor. During the war there was an opinion rather general among the people, that you surgeons often cut off hands, arms, feet and limbs very recklessly that might have been saved ; this does not seem to have been the case in your hospital, but what have you to say as to the impression ?"

I am very glad to answer this question. There can be no doubt that occasionally there was reason for such an impression in the hasty surgery of the battle-field. It was not so likely to occur in hospitals, however; it is human to err in judgment, but it has always been my impression that more lives were sacrificed because of attempts to save various members of the body than by members needlessly cut off. I lost several cases of belated amputation by reason of over-taxed vital power. There was one case of such intense interest that I think all lay readers will be glad to know about it.

A soldier was sent back to our hospital with a severe wound of the upper-third of one of his thighs, so severe that an amputation might well have been considered. When he reached us so much time had been lost that death seemed inevitable. To determine what should be done, even in such a hopeless case, his surgeon had him taken to the operating-room and we were all summoned for a consultation. On examination under anæsthetics the wound was found to be so extensive that an amputation would be necessary, but evidently so hopeless that he would be likely to die under the operation, such had been the exhaustion of vital power. The question was so close,—and really there seemed to be no question,—that a vote was actually taken as to whether he should live or die, and by a small majority the operation was decided upon—and the result was, a life was saved.

Readers, listeners, the surgery of the war was elevated to the highest possibilities of the science and art and the results were surprisingly successful—they seem more so now—the fact considered that we all were unaware of how our efforts were opposed by myriads of pestilent enemies, the bacteria of modern times.

After this rather elaborate unfolding of the advantages of such a hospital to the surgeon,—and what would elevate him professionally would, by as much, be of advantage to the patient,—I will now go back to my entrance upon my Herculean task, and tell you how I got through with it, for your curiosity was excited, not to say your interest, and no doubt your sympathy (for my patients).

To begin, I can assure you that my steps through the aisles of my ward on that first day were very deficient in the tone and firmness that confidence begets. But I had a right to believe that most of my sick would get well, and from my knowledge of the inwardness of prescription structure, and from the structure of my mind, I was simply incapable of becoming very apprehensive as to giving less drugs than nature would actually ask for, and nothing in authorized medical science was capable of adding to that apprehension. None of my patients got any dangerous doses on that day. But I found a great deal to do that I could easily and naturally do; there was mechanism in the dressing of wounds, in the coaptation and retention of the fractured ends of bones, there was force in the kind and encouraging word. Not one of my patients was aware that he had fallen into the hands of an amateur, and I never permitted one of them to become aware of the fact thereafter; not one was aware of the internal commotion over the much that must be done where all was so new; for, though in doubt over the materia medica, I was in no doubt but that more or less of drugs must be given, and *what, when, how*, were paralyzing questions that faced me before each of the sick.

Let me give you a "pointer" just here, or perhaps a reminder; whenever you are in need of knowledge in any important matter, ask the first person you meet for it, if you think he has a supply on hand, and without any regard as to who or what he is ; that is just what I did, and very often while I was learning to walk through my ward. My sources of supply in cases of urgent need were large and of high character.

It did not take me very long to get a degree of assurance that enabled me to go on with my duties with some ease, and with very little apprehension of being found professionally short by the surgeon in charge.

How did I treat my patients ? you naturally ask. Well, in a general way, as did all the other surgeons, except that the dosage was somewhat less vigorous ; and how could it otherwise be when on getting well acquainted with my fellow-surgeons and finding out their medical methods, it seemed to me *that cases of disease had a much more striking resemblance than the treatments to which they were subjected.* I was impressed by this, and because of it I could not believe that dosage was as important as was generally believed. I failed to see any science in treatments so diverse. *Science is always exact and never absurd.*

Having kept you with me in the army for some time, before mustering you out of the service, I will conclude by informing you that I remained on duty for a year and a half, during which my clinical advantages were far greater than I could profitably use. I never had any reason to believe that my services failed to reach a full average of success, professionally and in every other way ; but I must be frank. I must say that I failed to have a single case where I felt certain that a human life was saved by the use of the medicines given. The mystery of cure was as deep as ever.

LECTURE II.

My Friends the Readers :—

In my lecture of yesterday I failed to note that my materia medica suffered several abridgments at my hands while in the army ; and that the last abridgment by the exclusion of all forms of alcoholics, so reduced its bulk, that I easily brought it back in my vest-pocket. As I told you, I treated my soldier-patients in a general way as did my fellow medical officers, in the use of stimulants, but these were always used with a good deal of doubt whether they were not more harmful to the sick than they were to the well. I dosed my patients, then, as was usual, because I knew no other way ; but with less of energy and faith.

I have called you before me this morning to listen to a mere sketch of the history of eleven years of a general practice of medicine, that were years of groping, of reaching for a higher plane of service for the sick ; years of constant striving for clearer conceptions of Nature's power in disease : and so you will listen to my story with patience and perhaps with interest, void as it will be of heroes and heroines, of startling plots and surprises, and unrelieved by even a shade of romance.

I opened my office for this general practice and for this blindly groping of my way along beside Nature, very near her but always afar, in that my approach was at an angle very acute and therefore long, in the autumn of 1866, at Meadville, Pa., a city of ten thousand inhabitants.

There was no "felt want" to be supplied by this opening of a new office ; for the supply of able, experienced and well-read physicians, was in excess of the demand.   But this could not be taken into account, for there must be the "bread of life" from the first, even though it had to be gained in the form of crumbs.

My experience in the army had given me familiarity with wounds and diseases, of the severest character ; with death in its numerous forms, such as enabled me always to maintain my intellectual poise at a fair balance in the homes of the sick, no matter how severe the case.   A cool head in time of trial when others are unnerved is of vital importance, wherever anything vitally important is to be done.   This was one part of my stock in trade wherewith to support my sign, "Physician and Surgeon."   And I had become very familiar with the symptoms and history of the common forms of disease as they occur in men at their robust periods of life, and hence had become able to interpret symptoms and forecast probabilities with a little of that assurance of which much is required by the friends of the patient when there is an agony of apprehension.   But it was confined to this class of cases ; hence I was compelled to enter the homes of the sick with this one-sided experience, and a very small side at that.   I was very far from being a physician and surgeon in the sense of the gilt letters of the sign.

With this limited experience I was to go into the homes of the sick as an expert, to treat disease occurring at any age of life, patients of either sex, of every shade of constitutional strength and acquired weaknesses, and to treat any kind of a disease common to an inland city.

"Presumption," do you think ? That word expresses it. The general practice of medicine calls out and keeps in habitual exercise more of the resources of reason, of judgment, of philosophy than any other human avocation ; and there is no other avocation so taxing to vital power where the sensibilities are delicate, because you are expected to stay the icy hand of death even after it has begun to close upon its victim with its chilly grasp. You are not insensible, when you are the very center of the emotional tension that surrounds the bed of the severely sick,—you cannot be insensible to the emotional storm that surrounds the bed of death. And you are not insensible to your need of philosophy, when superstitious faith in your powers holds you responsible where death is inevitable.

This contact with emotional tension is a disabling experience, that confronts every physician at the bedside of acute disease, and at times it is paralyzing in its force.

Since I began the preparation of this lecture I have treated, successfully, (in partnership with Nature), a severe case of fever where the patient was a delicate young lady who had a father and mother, four sisters and three brothers whom I was to meet day after day with every glance of the eye, every expression of the countenance fairly vocal with the warning, " Have a care that your dosing is kept up fully to the need "—and had my case gone down to death, would it not have been an experience that would have cost me something of my own life ?

The care of a ward full of sick and wounded soldiers is child's play compared with the care of even one very

sick patient, surrounded by the family, friends, and
neighbors, who are morbidly or sympathetically interested.   This taxing of vital power was inevitable.

Another difficulty that I encountered at the very
threshold of practice was the *unreasonable, unreasoning, superstitious faith of the people in the power
of drugs to cure disease*, and it was not in the least
confined to classes.   Your college professor, your statesman, your lawyer who goes before juries to subject
evidence to ultimate analysis, will take your dose with
unquestioned faith when his life is in peril.   Your
railroad president, after surrounding himself with the
highest possibilities of home talent, will send his private car to another city for a medical expert whose
doses may, possibly, reach the point of disease with a
better aim.   This faith in the power of drugs is universal, and perhaps as strong to-day as ever, but is less
associated with the absurdities of that superstition
which characterized the earlier ages of medicine.
Modern superstition is of a more refined character with
the people and their physicians.

There was never a time in all history when chemistry and pharmacy were so busy in the manufacture
of new remedies for disease as they are to-day.   Our
medical journals are constantly adding to their pages
to make room for new and striking names, and already
our materia medica has become an encyclopedia.   Why
should we ever endure sickness when cures swift, unerring, are to be found on the pages of every news-
paper and magazine in all the land ?

With a painful sense of the difficulties to be encountered with every case of acute sickness, and of my defi-
ciencies from lack of experience, I began to receive
calls to exhibit my skill in cases of acute disease.   I
entered upon those services with the highest possible

professional pride in my success ; with the most earnest
desire to make those services effectual, by that closeness
of attention that was needed to make them satisfactory ;
and by all the cheer and hopefulness of manner that
needs to shine in every sick-room, where all tends to
darken life with gloom and apprehension : and my
sensibilities were such that I would naturally seek to
aid Nature by infusing all the cheer possible that is
always a necessity in and about a sick-bed.

I shall now try to make you understand, as nearly
as may be, the reason of my confusion over the question of dosage. I have already told you that most
cases of disease recover regardless of treatment. If one
of you has a watch to repair, go to what jeweler you
will there will be a striking similarity of the tools used ;
that is the science of mechanics. If one of you suffers
from headache or an attack of indigestion, and you go
to half-a-dozen physicians, and get as many prescriptions, then compare them, the chances will be that none
of the parts of one will be found in any of the others ;
that is the science of therapeutics. Very many diseases,
such as pneumonia, pleurisy, measles, etc., are very
much alike in all the features of their general history,
and always very much more alike than the treatments
to which they are subjected. I received a hint of this,
you remember, while in the army, and years of experience have only made it more evident.

Among my professional brethren whom I was to
know intimately for years were men of learning, of
high character, ability and experience ; and alike only
in faith in remedies. There were those of such exalted
faith and executive force that every patient was compelled to exhaust the resources of a small drug-store,
while the conflict between the disease and the remedies
was on. And there was the other extreme. One

of them informed me that he was treating a common disease with the 30th dilution of sulphur, and with good success. I scanned his expression closely to see if he were trying to measure my power of belief without evidence—but it was serenely sincere. Now his dose, of the same remedy, for the same disease, if it were even enlarged to the 10th potency, would be to mine, in size, as the microbe to the solid earth including all the inhabitants thereof! Now where, between these extremes, is the science of cure to be found? Or is it to be found all along the line? What think you?

Again, is not the very multiplicity of new remedies an impressive indication that the older ones have not proven satisfactory? And yet each in its time was supposed to be a veritable necessity. Drugs are subjected to chemical changes that are exact in every process after they reach the stomach; is it not possible that at times it is so dry, as in a case of fever attended with intense thirst, that it is unable to become changed into a chemical laboratory? Is it not unable, in fact, to perform chemical operations? In this light does dosage seem to have any right to be classed with the exact sciences?

Now, in all the small cities and towns, a physician hears more or less about most of the cases of severe sickness, particularly of those ending fatally, and very particularly where there are indications of an epidemic of a disease; hence some estimate can be formed of the comparative success of physicians. After close study of this matter for many years it is my strong conviction that the death-rate in a practice has a range in harmony with the number of patients treated, and is very little dependent on the science or the art of therapeutics, crude or refined.

My friend of the 30th potency had a large general

practice, and, for aught the public could see, he was as successful as my friend of shot, shell and canister, and both as successful as we who were between these extremes.

You will now better understand that either the science of cure has not been discovered, or that drugging the sick is a most haphazard adaptation of a means to an end. And you will also understand that when means so diverse are attended with results so nearly uniform you are often likely to exalt the physician far above his deserts, and to crucify him where death was inevitable.

Can you not easily see, then, that I was very likely, not to say certain, to have average success from my first case, no matter what my line of treatment?

Having raised serious doubts in your minds as to the science of dosage, you have some curiosity as to how I met the question in a practical way at the homes of the sick, where something has to be given every one, two or three hours, or a life is not being saved. As my lectures are not to tell you how, or when or with what to dose, but how to avoid, I may say little more than that I had from the beginning a strong point in my favor, and for the patient as well, in the matter of palatability. I had not been unimpressed by the fact that homeopathy had secured a very considerable foothold among the people, and it seemed to me to be largely due to its consideration for the sense of taste, reinforced by a success that seemed to me a full average beyond question. Certainly it would be so, if most cases of sickness recover regardless of treatment. My own sense of taste being very delicately adjusted : with a painful aversion to all things bitter, for sympathetic reasons as well as for weakness of faith, I was moved to make my doses of a character that would neither offend the sense of taste, nor structurally impair the membrane of the stomach, and therefore save my patients from wandering after other "gods."

The sense of taste is a very powerful factor in the matter of the nutrition of the body, as I shall show you in a future lecture. It was not made to be violated by strong dosage. Many of you have your ailments for which you have heavily dosed, and each dose taken never failed to cause this delicate sense to reveal its disapprobation, by a serio-comical expression of countenance; serious because painfully offended; comical, as if palliation for an absurd proceeding were being attempted. I will not have you believe, however, that some dosing is not necessary. We all are compelled to dose at times. We all use the needle in a case of agony, even my friends the homeopaths, and in its dose and its results may be found a close resemblance to exact science. You will understand, then, that it was a very easy matter to do all the dosing the credulity of the patient and the friends required, without encumbering nature by violating the sense of taste or the structural condition of the stomach.

With such conceptions, I began to visit the homes of the sick, and it was not so very long before this service became a specialty, almost in the sense that it would have been had I limited my services to one or more forms of disease, and without any design that it should be so. It came about in this way. I, myself, was somewhat the victim of that monster with many heads once called "dyspepsia," and had been for years; and even before I began my studies, homeopathy and allopathy each had its "fair and square" trial to deal with symptoms and not causes, but with total failure. Later, as a student and as a practitioner, I was ever on a search through libraries, for living vital knowledge that would relieve me of my enemy. I never found a remedy that was not to be pitted against a symptom, never a suggestion that was helpful by an aim at the cause, and why should I have so found? My enemy was in ambush, and I could not know his size and proportions, nor his power of resistance, nor even when or where to aim, and hence could not know whether to load with bird or grape-shot; whether with solid shot or shell. And then it did not seem wise in a matter of so much doubt to subject my sense of taste to a series of revolting and debilitating experiences; nor my stomach to the structural effects, more or less damaging, of doses that would excite the mouth into convulsive efforts to relieve itself.

The science of dosage had nothing for me then, nor has it now, nor has it for any who are suffering from the multiform ailings, due to deficient nutrition, through failure in the digestive process. If this is not true, why so many sanitariums; why this endless procession from one physician to another; why one side of every drug-store covered with patented cures; why their names painted in clearest lettering on every fence-board and on great rocks along the highway, on every iceberg of the polar sea? (If only the heavens could be painted what lettering there would be!)

No, I soon found I was not able to cure aching heads, nor neuralgia, nor diseased livers, nor palpitating hearts; that I had no appetizers whereby people could eat with relish; no pepsins that could pardon the sin of untimely feeding. For me, then, to attempt to cure with such conceptions, was to deliberately beguile revenues from failures believed to be inevitable; to decline these attempts was better morals, perhaps, than business sense, and so I let chronic cases go to other offices to encounter failure.

Soon after I began my practice I formed the acquaintance of a young physician who had built up the largest practice of the city ; that acquaintance developed into a lifelong friendship.   He had just begun to lay the foundation of a surgical career that was to make his name known the state over.   Very soon he began to call on me for assistance in his operations, and to make calls in an overflow, and for many years I cared for his business during his vacations and absences, and until a trained assistant became necessary.

Now his was a larger faith than mine, and of a much more forceful energy in medical therapeutics, and it was fully in accord with the methods authorized by medical science, and it was kept in such accord as the years went on ; hence I had abundant chances to compare results of our respective methods.   The great majority of his cases recovered, as did mine, and I never saw any such striking results from his stronger dosage, as to induce me to add to the strength of mine.

Confining myself almost exclusively to acute sickness, I used my little materia medica with all the care and discrimination I could master, for there was occasional need of the helping hand when Nature was prostrate, borne down from (too often) avoidable cause.   And I went on year after year, doing my utmost to achieve special success.   In a general way I dosed as I did in the army, but always with all the deference possible to the sense of taste, and of the integrity of the stomach, and I never permitted my homeopathic friends to drive me out of nurseries by their lighter medications.

As I gained experience in the history of disease, so I became more helpful to patients and friends ; and I was not so very long in finding out that the greater part of my care in every case was to minister to the need of the friends, and that this need can very often be duly met even in the most intelligent, by practitioners (not doctors), morally corrupt, mentally coarse, socially uncouth and professionally stupid. All that is required is *confidence,* and no matter whether founded on a rock or on the sands. In a matter of business importance, confidence must have a clearly defined foundation upon which to rest—but rarely in a case of sickness. Let me illustrate. A few weeks ago a patient fell into my hands, in the last stages of dropsy from disease of the heart. No encouragement was given; none could be given. Two men called, and with infinitely more zeal than sense, insisted that the services of an expert be obtained ; and the result, a self-constituted vendor of drugs who was also a farmer and an anti-prohibitionist. He could cure that disease, for he had done the like often, and even in worse cases, and hope revived in the invalid chair and through all the house. But death none the less occurred, and before a second inspiring visit could be made. Confidence, confidence, unreasonable, absurd ! Such experiences are common in every community, and are very far from being confined to the homes of the unintelligent.

As the years went on I believe my services became more and more valuable to all the needs, even to the minutest, which were revealed in and about the sick-bed except in the possible shortage and weakness in the dosage. Did I become a physician of marked success ? Scarcely. At times I thought I was certain to become such, and then would come a paralyzing experience. Why, I was called upon to take charge of four old physical wrecks at about the same time in close proximity in a suburb ; in a few days there were four dead bodies awaiting burial at the same time. To the vicinity there did not appear in all that a " specialism in success." Had that same experience occurred a few years later it would have excited notice, and derogatory comment, in all the city outside of my own attached families.

Other experiences occurred less marked, with sufficient frequency to keep me in a measure from the evil of over-confidence. But these are met by all physicians as well. What did I gain by eleven years' attendance upon the acutely sick, whereof no case was ever neglected, not even a visit missed because of disability myself ?

1. A great deal of experience, including that derived from nine years of service as physician to the county Infirmary, and something of that intellectual growth that must come where one mind comes into habitual contact with other minds of every shade of culture and power excited to vigorous action through fear or apprehension.

2. Only the reputation of being a painstaking, attentive physician, and a business created amply for the needs of professional culture, large enough to exceed the needs of daily living, but less, perhaps very much less, than would have been but for a presumed defect in dosage. *1.* It was too strong, apparently, to attract business from homeopathic considerations. *2.* It was too weak in a general way, presumably, to attract business from "allopathic" considerations.

And what the result upon myself ? A feeling that I had reached my highest possibilities in the science and art of treatment of the sick, and that in a business and a professional sense I had fallen far short of my ideal.

The mystery of cure was as deep as ever.

LECTURE III.

My Friends the Readers :—

I closed my lecture of yesterday with the suggestion that, both in a business and a professional sense, I had reached the end of eleven years of the closest attention to all the requirements of my practice in a rather disappointed state of mind.

One of the most discouraging things encountered during these years was the fact that a very large business could be done without any vital knowledge of the business to be performed, a fact, as I told you, I got a hint of behind the prescription case. This is really an anomaly in business affairs. That an intelligent, thinking, reflecting human being will put his own life or the life of his wife or child into the hands of those who, in every scientific sense, are as well fitted to perform the most complicated scientific work of any kind of business without previous knowledge or experience, as they are to outrage the human stomachs of the sick, is a marvel in human affairs.

As I have intimated before, this arises from the superstitious faith of the people in the power of dosage over disease, and this faith of course arises from ignorance.

Cheer of mind, as I shall tell you about later on, is such a prime necessity to health ; and disease, with its

47

associated possibilities, is so gloomy a subject to even think about; that the mind recoils from thought upon it; hence results the prevailing ignorance even in the most cultured classes, and hence the ease with which all classes alike become the prey of the spoiler in times of emotional tension when reason has become impaired, and strong words, and strong assurances are received as cold water by a thirsty soul, even with child-like faith, no matter how founded upon ignorance. This must ever be so until the people become aware that disease is largely a condition arising from avoidable causes, and that the cure is largely a matter of Nature's own handiwork.

I also told you that I had reached a condition of mind when there seemed no more progress for me—a condition that was not founded on satisfied ambition. I was in a fit condition for the incipient stage of a wilting; a dry rot of all the faculties necessary to my taxing duties. But this was not to be. On a hot day in July, 1877, I entered a home to assume charge of a case of fever that was to rouse every possible faculty called out by care of the sick as by an electric charge. I was now to have a revelation of Nature's power in disease, and that was to be all the more impressive because of the untoward conditions of environment.

The disease was typhoid fever; the patient a young married woman of rather full habit, that is, she was rather well rounded from undue proportion of fatty material, the muscle portion being light and weak. Her health had always been unstable, mental and moral faculties weak, and ambition scarcely large enough to furnish substantial reasons for being alive. The home consisted of one room and a small adjoining kitchen; this room was unprotected by a shade-tree to shut out the hot rays from a southern exposure, with no fly-screens, and the patient had only hap-hazard nursing.

Now I am not going to give you any history of symptoms as I would need to do if you were an audience of physicians. I will tell you that there was the foulest tongue I had ever seen in a sick-room, with an intense aversion to food. Now, to support the strength and vital power, it would have been my duty to enforce feeding in spite of the fact that the stomach was in as abnormal a condition as the tongue, and as functionally disabled. But Nature would have none of it. Every dose, every drink of water was instantly rejected for a period of three weeks, and it was not until the twenty-fifth day that suggestions of beef-tea failed to excite aversion. What was the condition of my patient? Was there a hopeless collapse of all the vital powers because they had not been adequately supported by the digestion of food by that very sick stomach?

I was a very surprised doctor, for, as the tolerant condition of the stomach was approached; even without food I found the tongue cleaning, and a manifest gain in both mental and physical strength, that became even marked at the time when food and doses could be borne; and I was moved to let Nature continue to have her own way. And, from thence on, I only watched, without enforced feeding, and with only unmedicated doses, until the thirty-fourth day, when she politely bowed me out.

This was an object lesson:

1. Vital power supported without food.

2. Mental and physical strength increasing with the decline of symptoms.

3. A cure without the aid of remedies, and a cure that was eminently complete in every way.

4. No unusual wasting of the body.

4

I was set to thinking, to reflecting as never before. I tried to recall all I could of every seriously sick case I ever had with reference to these four points.

I failed to tell you how I had fed my sick during these eleven years because I wished to reserve it until I could make it more impressive. I always permitted my sick to feed or be fed, when there was not special aversion, and with milk as the preference. But I never enforced this even in a single case.

It soon began to occur to me that the sum total of food taken by every severely sick case was infinitely small in proportion to the indicated need, and always too small to account, reasonably, for the support of vital power. And it began to occur to me that there was always a waste of the body in continuous operation, during the entire period of aversion or indifference to food. I had noticed this with eyesight but not with insight.

I had had numerous cases where for so long a time so little food had been borne that I might have wondered, but did not, how vital power was being supported. Such cases are constantly occurring in the practice of every physician. I began, too, to become aware that in every case, when the appetite point was reached, there had been a decided gain in mental strength that was not to be accounted for by the support of digested and assimilated food.

Is there ever a happier expression on all the earth than is revealed by the invalid who has reached the point of keen relish for his food, before the lines of expression have become drawn by the tension of his affairs? Is there not mental and physical energy revealed in every line?

Readers, listeners, you who have become in some

doubt as to the science of dosage, was there not enough of suggestiveness in this case to move me to try to find out whether Nature might not do as well in my very next case ? I was so moved, and I began a series of tests that were to be conducted with the scientific spirit and with scientific interest and accuracy.

Before giving you a thread of the history of my course of investigation, the following points naturally raised by the case of fever may be properly discussed, that you may better understand the logic by which conclusions are to be drawn, and therefore your estimate of Nature's power in disease may be more easily enlightened and deepened :

1. *Can the severely sick digest food ?*

If so, it is criminally stupid not to give it ; if not, it may be criminally stupid to give it ; if not needed, the disposal of it, without undergoing the process of digestion, is not only a needless tax on vital power but a most serious one, and by as much a hindrance to Nature that may often turn the balance in a fatal direction.

There are as many methods in medicine as there are doctors in medicine ; each doctor is largely a law to himself as to the size and frequency of his doses, and, no matter how closely diseases may resemble each other, there is rarely any uniformity in either general or special treatments, as I have already told. But no matter how doctors may differ in methods and dosage, there is a very general concurrence of opinion that the aversion to food that characterizes all cases of acute disease, which is fully in proportion to the severity of symptoms, is one of Nature's blunders that requires the intervention of art, and hence enforced feeding regardless of aversion, or of conditions affecting digestive power adversely, is nearly universal. And it may be stated as a general fact that the desire to support vital power by feeding, is in proportion to the apparent need, by reason of the severity of the symptoms.

In the consideration of this vast question as to whether the sick can digest food, some suggestive ideas may be derived by a study of the conditions of normal digestion as they exist in their highest type, in contrast to the conditions as they exist in disease.

We find the conditions in their highest type in childhood and youth. And the first and most striking condition to be noticed is the continuous cheerfulness of mind.

The child, the youth mind seems to be in a state of continuous ecstasy which is rarely depressed unless by a task or disease. This cheer is to digestive power, what the draft is to the flame, and so continuous are the demands made upon the blood for sustaining and building material that there cannot be the least depression of mental cheer without cutting down the supply, by lessening, the very motive power of the digestive function, and causing, not seeming, but actual fatigue or loss of mental and physical power.

We may see an exhibition of this needed cheer in an exaggerated degree in the yard of the ward school during recess, where pent-up Nature finds relief in the tumultuous action of muscle and mind.

There is such running and jumping, such screaming and laughter, such acuteness and intensity of mental action as to suggest that an unseen hand is at work to restore a lost balance.

Cheer of mind is a primal law of life ; it determines what every human employment shall be where there is power of selection ; it is the very motive power of all human effort, whether of virtue or vice.

There is no period of life when the digestive power needs the support of mental cheer as when the foundations are being built up ; no period when Nature guards this source of power with such unceasing vigilance.

Then there are those general muscular activities in the fresh air that cause the keenest of appetites for the most nourishing of foods, and also that sense of general fatigue that makes the bed only less a luxury than the playground, and sleep so perfect as to scarcely interlude the days.

In disease or injury all these conditions are reversed, and, for an illustration, the case of President Garfield may be cited. He had reached the highest pinnacle of his human ambition, and when he was on his way to a joyous reunion with his fellow alumni at the home of the alma mater, he might well have been considered the happiest of all men ; but he was overtaken by the assassin's bullet and laid low.

His extensive knowledge of wounds from his army-experience required no hint from surgical authority that he was under sentence of death, and from thence on there was no cheer in life for him. He never saw the countenances of his great surgeons illumined with a single ray of confidence ; the severe exactness with which every detail of his care and treatment was carried out was continuously ominous of the gravity of his condition. The frequent taking of pulse and temperature to satisfy the greedy wants of the bulletin-board was a long series of blasted hopes. That there was a painful perseverance and rigidity in all that was done for him ; that there was only continuous doing with drawn countenances and hushed manners was in-

evitable, for he was the center of the intense interest
and sympathy of the world.

With even the slightest shade of gloom having its
degree of depression upon the digestive function,
was there not enough in the president's mental condition to nearly, if not quite, paralyze this function?

The bullet tore its way through several inches of his
body, leaving behind a track of lacerated nerves. The
agony from his wound soon began to be reinforced by
the agony of a slowly developing abscess, by which
tender nerves were kept in a continually increasing
tension. From thence he was never free from pain
except by the mercy of drugs.

With the slightest shade of discomfort having its
degree of depressing effect upon digestive function,
was there not enough of agony in the president's case
to nearly, if not quite, paralyze this function?

To maintain normal digestive power there must be
more or less muscular activity in the fresh air, reinforced by perfect sleep.

The president's body had to be kept motionless by
reason of his wound. There was no breath of life in
the stifling air of the White House, not one moment
of natural and therefore invigorating sleep. Was
there not enough depression in all this to nearly, if not
quite, paralyze digestive function?

There was continuously a high pulse and temperature; the mouth, the stomach, the entire digestive
tract was therefore abnormally dry; the fountains of
the saliva and gastric juices were nearly if not quite
dry. Was there not enough in this to nearly, if not
quite, paralyze digestive function?

*It is Nature's plan that digestion shall be performed
after the general muscular activities have created a*

demand for it, and with the body in the erect posture ;
this posture favors the somewhat circular sweep of the
food around the walls of the stomach, known as peristaltic motion. In the lying posture this process in mechanics is imperfect ; the dependent portions of the
stomach being somewhat unduly affected by the weight
of the food from the force of gravity, and then, in the
lying posture, particularly during sleep, there is so
little waste of tissue from cell destruction, that there
is very little if any need of digestive energy. There
was excessive general prostration of muscle energy,
which was shared by the muscle fibers of the stomach,
hence enfeebled if not paralyzed peristaltic power.

Was there not enough in this and the continuously
prostrate position to seriously affect digestive function ?

In addition to all these depressing forces that were
inevitable there was another of great power that was
inevitable because strongly indicated, and that was
the medicine used to relieve pain and enforce sleep
(stupor).

To make you see clearly how such medicines paralyze digestive power I will illustrate by a case drawn
from real life, and that will not be tinged by a shade
of fiction.

I was called to a home where I found a patient that
had been confined to a bed for nearly a month as the
result of an attack of cholera morbus. The treatment
that had been instituted was, naturally and properly,
the hypodermic needle to relieve the agony. Now this
so-named attack might possibly have been called an
insurrection, a convulsive commotion, or a battle in
the stomach and bowels, the " tug of war " being over
a food mass that had to be disposed of without digestion ; or it might have been called an operation in

vital chemistry, whereof the new combinations were natural, but none the less inconvenient to the patient who had to furnish the laboratory ; but the dose was indicated.

To keep up the strength, and therefore to support vital power, a tumbler of rich Jersey milk was taken within three hours after the dose was given. Now it is just barely possible that the so-named disease was only a chemical commotion, badly located and wholly due to the taking of food not wanted, and that therefore should not have been taken.

Opium and its derivative, morphine, say the books, in addition to power to relieve pain, also "*dries the secretions.*" Well, the saliva and the gastric juice certainly are very important in cases of digestion. How then is milk to be digested, if the foundations of these secretions have become dry ?

We are not enjoined by these same books, nor by any other authorities, for this reason, not to give food while the secretory glands are in the drying grasp of this powerfully paralyzing drug ; at least such a hint has escaped my notice, if to be found, as it has every one of the many physicians to whom I have suggested the matter. The giving of the milk, then, in connection with the dose was accepted medical science. And it had the support of so learned and able authority as Prof. H. C. Wood, who tells us in his Practice of Medicine that in such cases as pneumonia, scarlet fever, etc., we must not withhold food beyond the second day, that we must feed even if there is decided aversion to food ; and in cases where the stomach has become so disabled that none can be borne, we must resort to nutrient enemas, and that if the temporized stomach proves restive then we are to throttle with a quieting dose of laudanum mingled with the food !

But how efficient could be the dry excretory membrane, and the paralyzed muscle tissue beneath it as an aid, a vehicle for the transportation of food so digested as to be ready to fairly drop into the form of living tissue? The professor has great faith in this means of supporting vital power where the stomach is disabled, and he cites a German authority who, in a case of total disability of the stomach, kept his patient alive a whole month—remember the time, readers—*a whole month.*

It is barely possible that laudanum and food, no matter how predigested, might not form a good combination for digestive and assimilative purposes when taken into the stomach, nor would the sense of relish be likely to be restored when the aversion is due to disease.

The tumbler of milk duly taken, and others later on, either because there was a recurrence of the attack of disease, or of another conflict over cheese curds, the needle was again summoned. Now, for nearly a month there were taken from three to five tumblers of this rich milk during the day and the night as well, and more than as many doses of the "quietus," and all the while a loss of strength and a wasting of the body went on. There was just a bare possibility that there was no digestion going on the while, for the foregoing reasons.

Now, my treatment involved a radical revolution, a challenge, in fact, to be flung at the very teeth of authorized medical science.

"Madam, you are to take not one drop more of milk, not another dose of morphine. You will have a restless night, but when the present invoice of cheese curds are disposed of you will have no more pain. You are going to let that stomach rest until it calls as loudly for food as it did for relief. You will not starve ; on the contrary, you will so gain in strength right along that it will become very marked by the time you get hungry. And mark you, when real hunger comes, when you let Nature rest until she can so speak that you can hear every word she says, then give her the exact bill of fare she calls for, and no mistake will be made."

Result. Second day.—A restless night, but on the whole stronger and more comfortable. The vital powers had now become freed from the grasp of the last dose, and there was repose all along the line of recent conflict. No hunger. Third day.—A better night with increased mental and physical strength. There was an attack of hunger on at 10:30 A. M., and nothing but a boiled potato was called for, and it was taken with such unction of relish that a second meal of the same was required later in the day. Fourth day.—Still improved and bill of fare enlarged. Fifth day.—Light duty begun and my services no longer needed. My *dosage* had nothing to do with the cure in this case.

Now, readers, is it not barely possible that this was not one of those cases of a " fool for luck " ? Why, it happened some three years ago, and on a visit to get the points of the case freshened, she remarked, "I never in all my life ate anything that so refreshed me as did those first two meals of boiled potatoes." Nature generally knows what food she wants when she has a voice in the matter.

The treatment of my predecessor was very much more in accord with accepted medical science than was mine. Believing as he did that vital power must be supported by the taking into the stomach of a certain amount of food daily, he gave the only kind that could be taken without such special aversion as to nearly be prohibitory.

It has been my impression that we of the army sometimes dangerously erred in this way. I well remember a case where a great operation was successfully and skillfully performed, and the resulting wound completely healed, but the relieving dosage went on as did the three daily meals. There was no disease of the digestive tract, and it was specially indifferent to the ordinary irritations of decomposing food, and under the powerful grasp of the drug, there was kept up continuously an apathetic indifference to all things earthly. The position on the back was very nearly motionless, and nothing seemed to go on but seeing and breathing, and yet the three daily meals were kept up against these intensely adverse conditions of digestion, the body wasting away with every heart-beat ; but when the skeleton condition was reached, weeks after, death came, and perhaps it is barely possible there was a martyr to accepted medical science.

A few years ago, while on a visit to a sister in a city in the West, I had need to deliver a lecture on the dietetics of infancy by reason of the presence of a badly nourished infant from badly adjusted feedings. This led to a call to see an infant supposed to be near its death from some wasting disease. The doctors had "given it up," and well they might, for the skeleton condition had about arrived, and hope was gone. What was the trouble ? A great many irregular feedings per day and a great many doses to stifle Nature's protests. It was again milk and opiates, with the stomach and bowels the line of battle.

The theory was explained to the receptive mind of the mother of the first-born, a higher method of living

instantly adopted, and that mother to this day believes a life was saved thereby.

Readers, I recently read that we were never to try
to reason a prejudice out of any person's mind, because it never got into his mind by reason.   And now
let me ask you, you in particular, because the very
structure of your minds makes a prejudice impossible,
let me ask you if you can possibly believe that against
the array of adverse digestive conditions, the president
ought to have been fed ? He was fed until Nature
very promptly decided that the surgeons were wrong,
by permanently locking the stomach against all their
unbidden efforts to support vital power in the usual
way.   And from thence on, this power was to be supported through a less defensive portion of the digestive tract by means of predigested food.   Was it
possible to support vital power in this way ?   Let us
consider :

The stomach was functionally powerless, not actually diseased, and as it was one of the results of dis-
turbed vital power, the entire digestive tract was
probably in a similar powerless condition.

The mucous membrane of the extreme portion of
the digestive tract seems to exist for the sole purpose
of generating and throwing out a slimy coating for
the fœcal contents that protects by its antiseptic quality,
and renders their expulsion easier.   It is then an excreting, and not an absorbing membrane, though it is
capable of absorbing water and alcohol and medicines
clearly soluble in either.   It is therefore something of
a draft on credulity to suppose that nitrogenous food
could be made so soluble, as to be made capable of absorption by a membrane whose natural function is in
utter reverse to that of the stomach, and which must
have been very nearly as powerless as was the stomach.

The stomach is the firebox of the human locomotive, and there is no other that can consume life-maintaining fuel.

The mucous lining of the stomach is both an excretory and an absorbing membrane; it throws out its marvelous solvent juices when hunger indicates the need. There is instant digestion, instant absorption, and an instant and continuous sense of refreshment until hunger is sated, which is one of the most acutely pleasurable in all human experience.

No sense of refreshment ever comes when food is taken, even in health, without hunger, and because there is little digestion, little absorption; much less can there be in disease.

No one ever thinks of eating if the appetite is abolished by a *trivial ailment* and plainly for the reason that it would be an unpleasant experience attended by depressing results; but if the ailment is thought dangerous, why, then the physics and chemistry of digestion are utterly ignored and food must be enforced.

Was it not asking a great deal for the mucous lining of a receptacle, that was situated nearly thirty feet from the fire-box, or from the chemical laboratory which is our only reliance in time of health for the conversion of nitrogenous food into a soluble, absorbable condition whereby we live: was it not asking even a miracle of this excreting membrane which must have been as functionally disabled as was the stomach, to take upon itself the disposition of food artificially digested?

I see that you begin to look puzzled, for you have always believed that the president's life was maintained during those long weeks through the absorption of food by this most accommodating of excretory membranes, of food so marvelously predigested as to be ready to fairly drop into the form of living cells. I will add to your confusion. It is my strong conviction that *the president's life was not prolonged one second thereby,* and that the wholly unnatural and seemingly unreasonable method of attempting to support vital power was a torturing experience that was void of any credit to medical science.

I see you begin to wonder: " What you, you call in question any method of these great surgeons!" Great they were in ability, in skill, in experience ; great in all that ennobles and elevates human character, and they were the greatest of living surgeons, but——

One of the greatest of College Presidents once said to his class, " Young men, cherish your own concep- tions." in other words, do your own thinking and draw your own conclusions.

LECTURE IV.

My Friends the Readers :—

Before beginning my lecture this morning I will read a note I find upon my table written by one of you.

Doctor : As a matter of mere theory, you have made it strongly evident, that the president had no power to digest food, but you know how often it occurs that the best constructed theory has to go down before an inexorable fact. Now there are three rather striking facts in this case. 1. You admit that the excretory membrane had some power of absorption. 2. Predigested food was habitually used. 3. The president lived 80 days. Do you then expect us to believe that the membrane did not absorb and transmit enough of the digested food to prolong his life, far beyond what it otherwise would have been? Why, think of it, 80 days are nearly three months, and to believe that any human being could live so long without nourishment of some kind, taken in somewhere, is, to say the least, incredible, unbelievable in fact.

Sincerely your friend,
A LISTENER.

I am very glad to meet this question so clearly stated and so pointedly put, as the point raised is now to be considered under the second question.

2. *Is it necessary to feed the sick?*

The president was shot on the morning of July 2d, and I shall never forget the tolling of the bells near midnight of September 18th announcing his death, in tones muffled with a seeming grief deeper than human. What kept him alive all these weeks, months even, of agony? Food, do you say, artificially digested and most artificially administered, not to say unscientific, in conception?

I now draw for your consideration a very suggestive parallel. On the morning of June 4, 1886, I was called to see a frail boy of four years, of unusually spare build, who had just taken a drink of a solution of caustic potash. He was the only son of intimate personal friends and I had been attached to him from his birth ; and so all the more I was moved to watch his case with exceeding closeness and to do all that could be done to alleviate, and to save.

The solution that was gauged to clean the woodwork of the house was strong enough to destroy the stomach, and from thence on every drink of water, was ejected. Of course you will understand that drugs and liquid foods, would be even more objectionable to this inflamed stomach. It was not within human possibility to treat the inflamed stomach with any form of drugs, to support vital power with food predigested or otherwise. He promptly ejected the first drink of water, and as promptly the last he ever took.

I saw him draw his last breath, and it came seemingly from a skeleton covered by a thin skin, and this last breath was drawn on the morning of the 17th of August, 75 days after the fatal draught. The mental condition was clear to the last hour.

All that seemed to be left of the great president when he drew his last breath on the night of the 80th day at Elberon was a thin skin covering a skeleton. What became of the tissues in these cases?

Did they evaporate?

Now there is room for the suggestion, that if some predigested food could have been absorbed by the excretory membrane, it might have given the stomach a little more time for a possible merging into a retentive condition for some light aliment. But the sensitive membrane of the delicate child would soon have become so restive under the abnormal service, as to have required the subjugation of doses of laudanum which, mixed with food predigested or otherwise, forms a combination no matter how sanctioned by high authority, and which would not work well in the stomach, in the case of a picnic dinner.

To get into this question a little deeper. I am going to show you a table which I have taken from Yeo's physiology, an authority unquestionable. It shows the estimated loss of the several tissues of the body in cases of starvation.

FAT	91 per cent.
MUSCLE	30 per cent.
LIVER	56 per cent.
SPLEEN	63 per cent.
BLOOD	17 per cent.
Nerve Centers	0 !!!!!!!!

Listeners, there is a combination of words that when I first saw it, impressed me, pleased me, far beyond my power over written or vocal language to express. At last I had found a clearly-defined bill of fare for the sick, and a bill that I had been relying wholly upon for years before I saw it, and without any conception of just how vital power was supported in the absence of digestive power. A mere glance at this table seemed

5

to me to settle this whole tremendously important and interesting question of nutrition in disease.

I had never had any thought or conception as to how the integrity of the brain and nerve centers was provided for in acute disease. I had seen, as I told you, while in the army, that no matter how emaciated the body might be, the brain, the heart, the lungs never revealed any loss, but there was no seeing with insight.

The brain is the organ of the mind. the home of the soul : it is the power-house of the human plant. It is the source of all physical, mental and moral energy. Shooting out from its depths upwards, outwards, downwards, are its bundles of electric cords to divide into a vast net-work of single wires, with the sensation wires ending each at a little station, a signal station. How useful, and, at the same time, how like a structure of human devising does this battery operate !

An unbidden meal is eaten, and in due time there is set up the chemistry of decomposition in the temporized laboratory—a signal from the sensation station goes to the battery, and forthwith there are groanings that cannot be so uttered as to do justice to the commotion, the emotion within, in the appeal for aid.

And then there is the electric cord with its reinforcing batteries that presides over and regulates the mechanism of the vital organs, slow in its action but vital in importance. Seeing, tasting, smelling, feeling, love. hatred. happiness supreme, misery that drives to suicide, comfort indescribable, agony unendurable, all, all depending on the continuous integrity of the central *battery, the brain !*

Readers, is it conceivable that an organ so vital to life and to all that is worth living for, should be left to the

hap-hazard feeding of the stomach of a sick body?
" Tell it not in Gath."

Let us look at this table again, and first we notice
that 91 per cent. of fat disappears in the starved. That
is a matter of business sense on the start, if it is drawn
upon as nourishment by the brain, for how easily we
may get it back again, and what vast multitudes there
are who are heavily inconvenienced by vast masses of
it, that is sheer dead weight to be carried about with
hurried respirations. Now, if this is nourishment for
the brain and always available, need we be in hot haste
to force food into unwilling stomachs, as soon as the
third day of a severe attack of pneumonia or fever, if
also disease is also a depressing force to digestion?
We can always spare fat, when there is excess, without
loss of strength, as you know. The prize-fighter has
to go into training that the fat may be brought down
and the muscles brought up to the "fighting weight."

The president had a vast store-house of fat, my
starved boy a very small one, and yet how long they
lived without food ! With an ample supply of fat for
weeks of fasting, if it may be considered a store of
predigested food, is it reasonable, is it conceivable that
in the light of this table, even without other evidence,
the excretory membrane of a receptacle can be changed
into an absorber and transporter of artificially digested
food, when its irritable aversion to the service has to
be stilled by laudanum? "Tell it not," don't even
whisper it in "Gath."

The muscles lose 30 per cent.; well, we can easily
get that loss back when digestive power is restored,
and then we do not have so very much occasion to
use the voluntary muscles when we are sick. The
liver and spleen seem to be heavily drawn upon, but

we are never aware of special trouble from them on this account in time of sickness.

The blood loses 17 per cent. Perhaps the brain, also the heart and lungs, which are not down as suffering a loss in starvation, can be kept nourished with this per cent. below the normal of nourishing power, or perhaps it may be in a normal condition for the required degree of service.

Now, with this table right before you let me reiterate, a favor I shall ask again and often, for I cannot meet the purpose of these lectures without.

1. Nature takes the appetite away always with an emphasis equal to the severity of the disease, and with special emphasis if the digestive tract is the center of attack.

2. There is the progressive loss, or wasting that goes on unhindered by any kind of feeding, while aversion to food exists. In this connection I will inform you that some four years ago, when I had become as fully satisfied as I am now, that the stomachs of the sick are unable to digest food, and, without having seen this table, had, by experience, arrived at the conviction that vital power was better sustained without food than with it, I was moved to write a short article for publication on "Feeding the Sick," which contained this statement, and also the suggestion that there is an increase of strength, in accord with the decline of symptoms, in all acute attacks that recover, that becomes marked even before any food is taken. This article contained these, to me, original ideas, as well as some others, and where should I offer it for publication where it would not be buried alive? There was only one paper even for a moment considered. The "Scientific American" has more readers in the ripe maturity of their powers, and of discriminating in-

tellectual force, than any other paper ever published, and therefore the best paper in the world to put a new, live idea into.

To that paper I offered my first-born, a mere condensed outline of a much longer article that I had been long at work over, expecting that it might be worthy of the column of some high-class monthly. It was promptly accepted, published, and was widely copied. For more than two years it floated among the papers as a thing of life, or a derelict. It attracted the attention of an assistant surgeon of the U. S. N., a stranger, who joined issue with me on the question of progressive waste regardless of all feeding. He admitted the waste, but was very certain that, by the use of various foods, predigested and otherwise, this waste could be somewhat lessened. Of course he could not prove that such was the fact, nor I that it was not.

But mine was a much larger experience than his. It was not a question that could be solved by science. There could only be inferences drawn from a preponderance of evidence. Several letters were exchanged, and he was moved to write that " friction of minds similarly bent is useful to both," as the result on his side.

Now I was moved to go into print at that time, because an unusual prevalence of that intolerable disease, La Grippe, was moving a great many physicians to go into print on the importance of keeping vital action in support, by doing all the feeding possible ; the stomach must be kept busy ; the medical journals and newspapers were full of this sort of thing. And now I will draw a third suggestive parallel.

About this time a case fell into my hands of this same disease. The patient was a lady of very spare build, frail, of mature age, and had always been very much of an invalid ; in fact she had not known anything about vigorous health since she was a child.

I found her prostrate and in such apparent exhaustion, as to be scarcely able to whisper ; there did not seem to be power to even draw the breath of life. The pulse was rapid and even thready, and in every way vital power seemed so utterly prostrate, as to make recovery a matter of exceeding doubt. Now, these same physicians would in this case have forced unbidden food, their own bill of fare, into the stomach, with not one thought that it was as functionally disabled as were all other powers ; with not one thought that it would have been an additional tax on vital force ; great in proportion to the prostration already existing.

I am going to pause long enough for a suggestion that I will utter in Italics, and I want you to receive and blaze it into your memories in Italics, to be kept side by side with the table I have given you which you are always to remember as "Nature's bill of fare" for the sick. The suggestion that *when death occurs before the skeleton condition is reached it is always due to old age or some form of disease or injury, and not to starvation.*

Now, there could be no feeding in this case, for there was a continuous degree of nausea that would scarcely abide water ; much less any form of food or drugs. For thirty-three days no human doctor could have fed this patient because there was nausea every day, and scarcely a day without several spells of vomiting. What did I do the while—what could I do with a stomach that would not hold still long enough to let me try to cure (?) it ?

During the war, when the Proclamation of Freedom was pending, Gov. Richard Yates, the lifelong friend

of President Lincoln, became very impatient over the
delay, and to one of his pressing importunities, he received the telegram, "Stand by, Dick, and see the
salvation of the Lord!" And so I had to stand by
day after day as a minister to the need of the friends
and utterly unable to do one thing for the patient, but
with a keen eye to the wasting of the tissues ; to the
change of the mental and physical condition that went
ceaselessly on.

The extreme prostration of the first few days soon
became relieved little by little; the heart slowed down to
firmer beats, and all along the line, except the stomach,
there was a slight improvement, and the only question
with me was, will the bill of fare hold out until the
unknowable morbid condition behind that sick stomach
can be cured by Nature's divine hand ?

Thirty-three days of nausea, of starving, when the
patient, from thinness and mal-nutrition was half-
starved at the inception of the attack ! That was a
long time for the vital powers to go without support,
was it not, really ?

You will remember that the German physician actually proved that his case was kept alive a whole
month, by the kindly accommodation of the excretory
membrane; how would his logic fit a case like this ?

The thirty-fourth day came, and with it a stomach in
repose, and with every mental and physical energy
elevated to newness of life, but the body was exceedingly wasted. The thirty-fifth day came and with it
the "Salvation of the Lord," clearly, unmistakably
achieved, for there was a call for food of the same
digestive worth that must be on the noon-table of the
logger's camp of a January day in the Maine woods !
Nature had changed the bill of fare : the skeleton condition was not nearly reached, and a case of La Grippe

cured without that meaningless, that senseless, that
most dangerous stuffing that is warranted by accepted
medical science.

Will you kindly listen while I draw a fourth parallel ?   I was called to a case of typhoid fever after it
was under full headway, in the case of a frail girl
nearing her 14th year—a bad age for a case of fever.
A homeless orphan, she became a county charge and
was taken to the hospital.   Now the matron, a grad-
uate of a Woman's Training School, and well experienced from the care of patients under the masters of
medical science of New York and Philadelphia, had
heard a great deal about my shortness on the food
question in cases of the severely sick, and hence was
anxious to see a case, especially of fever, treated with
the inevitable milk and the commonly inevitable
stimulants absolutely prohibited.   And to gratify her,
and just possibly from considerations of humanity, I
retained charge of the case that rightfully belonged to
the paid service of the city physician.   And so day
after day, under the best nursing the case went on—
the doses, with rare exceptions, unmedicated, were
administered with exacting punctuality, and the fever
chart was made to reveal the nightly and daily changes
with habitual regularity.

And here I will pause to interpret another suggestion—that most of the need of drugs to allay restlessness and pain, and to enforce sleep in cases of the
severely sick, arises from the exhaustive taxing of
vital power from the enforced feeding and stimulation,
that is, I believed such to be the fact.

In this case there was a degree of mental apathy, I
thought might be normal, as the girl was a stranger ;
but after two weeks the digestive tract all along the
line became quiescent, and the mental condition had

become clearer and stronger, and an early recovery seemed assured; but one day she asked for food, and not being at all ready for it none was given, and a crying spell, as I learned the next day, resulted. Now, what could I do? The danger line seemed not yet fully passed, and to deny a crying call for food, would not do in the present stage of civilization. I felt that I positively knew it could not be digested, and yet as I could not actually know that recovery was assured, what would be the result to me personally were death to occur with the unheeded *cry for food*, to ring in the *public* ear? No, no, that would not do, and then it is often well to gratify the whims of the sick, if it can be done safely, when reason cannot sway.

I left my sick girl with directions to give a light feeding in case another attack of want should occur. It occurred, and on my evening visit the temperature had taken the highest jump of all of the days of the sickness, and an irritability of the entire digestive tract resulted that cost a week to recover from.

That settled the food question while the disease had sway; there was none called for again until about the end of the fifth week, when there was a clean, moist soft tongue and stomach as well, and food was wanted well worth the digestive process. The skeleton condition was yet afar off. And from thence on no wood-chopper ever wanted stronger food, nor relished it with a more acutely ravishing sense of taste.

The method of the German physician could not have availed in this case, because of the irritability of the entire digestive tract; and yet how would he account for the support of vital power for so many as five or perhaps six weeks, as the girl had been ailing for more than a week before I was called?

In this case the cure was one of Nature's own "clear-

cut " jobs all the way through, for such was the delicacy of the stomach that water very often caused
nausea, and much more, strong doses would have been
as objectionable. Can you believe that my part in
this transaction was a matter of intellectual stupor? If
so, do not, not just yet, " publish it in the streets of Askelon," for I am not yet done with parallels. I must
give you another and a very striking one. And in this
case, and in all that are to be mentioned, all that have
been mentioned, you are not to abate one "jot," not
even one slight " tittle " from the apparent meaning of
my words, for I am not unfolding my own but Nature's
power. I am making no appeal to your credulity.
Some of you have gone to the spoiler for aid in your
time of need, and by wise looks and strong statements
in language, you could not fully understand, your
hopes were raised only to be dashed. I have no
mystery of treatments to offer in words of "learned
length and thundering sound," no hopes to raise
other than founded on all the reason I can command,
and, where reason fails to meet your idea, then don't
make any effort to believe.

My next parallel will be drawn in the next lecture,
and in the meantime fix all the points thus far raised
firmly in your minds so that you can give each a careful, a deliberate consideration, before they can become
your own, or be rejected because they are not able to
stand the test of reason.

LECTURE V.

My Friends the Readers :—

The difference between one mind and another is often the difference between the machine-shop and the ware-room. One mind is all machine like a thrasher ; the sheaf goes in, the chaff flies, and the winnowed grain flows in a stream to meet the world's need.

Another mind is an absorbent with a ware-room annex, into which floats all sorts of odds and ends, to settle into a fixed, conglomerate mass, as deposits form the crust of the earth. And there is still another mind with its annex, an Academy of Fine Arts, from which nothing ever gets out, and into which no one can enter but the proprietor.

There was a Halleck and a Grant in war-time. The former cultured, stately, impressive, of large head and learned beyond any general of his nation. One day a genius, in a moment of inspiration, called him "old brains," and instantly a nation held its breath to see the rebellion go down with a crash.

There was the "silent man," unimpressive in every way, with his faculties asleep or withered from years of inaction ; or they were, perhaps, nuggets of gold, only waiting discovery, or perhaps giants slumbering in the most healthful repose, only awaiting the electric touch of a great opportunity to rouse them to powerful action.

75

Artemus Ward, the inimitable, one day said, " Some of our generals ought to be old women : some of them are." What was an encyclopedia when the supremest need of a nation was *"action, action, God-like action ?"* One a splendid figure-head perched upon a magnificent ware-room, and the other—Appomattox ! Grant was not the old woman.

The most learned physician I ever knew, learned because he read everything, learned because he never forgot anything, had no machine behind his encyclopedia charged with moral energy. His was a storehouse only, and he died, and no one was ever the wiser or the better because he lived. This same inimitable Artemus once said that "a well-balanced mind, is a mind that always balances in any direction the public requires."

Perhaps the most striking example of an immense ware-room filled to bursting with an assorted, classified list of wares, each ready for instant use, and all adequately supported with an engine charged with moral force, was revealed in the case of Franklin. Poet, philosopher, moralist, humorist, scientist, electrician, politician, statesman, he became the world's greatest schoolmaster, and, with his intense practicality, he ever used his vast stores, and rare powers for the good of all nations and peoples, finding therein his highest happiness.

And there was a Lincoln, surrounded with a cabinet ; all with infinitely larger ware-rooms, yet he was the intellectual superior of each ; unapproachable in scope, clearness and power of philosophic vision ; towering above them all physically, they were as children in the presence of his overshadowing, moral presence. No member of his cabinet was ever magnetically affected by the moral force of any of the others, because none "had rather be right than be president;" at least no one of them could approach him in such utter abandonment of self when a need was to be met.

Listeners, we ought always to keep our eyes open for truths, that we may be able to seize upon them, even if they eject us from our cherished beliefs or our beaten paths of life. To refuse to consider truth because not popular; because it seems averse to the even tenor of our ways of life is to condemn without evidence. One of my very best friends has never even permitted me to hint to him, that there is any physiology behind my "science of cure," such is his attachment to the beaten paths of medicine.

I am now going to give you an illustration of the very highest character, of the scientific mode, and persistence in the search for truth. I find it in a lecture on Scientific Culture by Prof. J. P. Cooke of Harvard, the first of a series on scientific subjects that you lovers of science ought to read. It is an account of an investigation of the composition of water by Sir Humphrey Davy. Sir Humphrey was a plodding old gentleman, very slow, indeed, to believe without evidence—was always averse to drawing conclusions by the aid of his imagination.

" The voltaic battery which works our telegraphs, was invented by Volta in 1800; and later, during the same year, it was discovered in London, by Nicholson and Carlisle, that this remarkable instrument had the power of decomposing water. These physicists at once recognized, that the chief products of the action of the battery on water, were hydrogen and oxygen gases, thus confirming the results of Cavendish, who, in 1781, had obtained water, by combining these elementary substances; oxygen having been previously discovered in 1775, and hydrogen at least as early as 1766. It was, however, very soon also observed, that there were always formed by the action of the battery on water, besides these aeriform products, an alkali and an acid ; the alkali collecting around the negative pole, and the acid around the positive pole of the electrical combination. In regard to the nature of this acid and alkali, there was the greatest difference of opinion among the early experimenters on this subject. Cruickshanks supposed that the acid was nitrous acid, and the alkali ammonia. Desormes, a French chemist, attempted to prove that the acid was muriatic acid ; while Brugnatelli asserted that a new and peculiar acid was formed, which he called the electric acid.

"It was in this state of the question that Sir Humphrey Davy began his investigation. From the analogies of chemical science, as well as from the previous experiments of Cavendish and Lavoisier, he was persuaded that water consisted solely of oxygen and hydrogen gases, and that the acid and alkali were merely adventitious products. This was undoubtedly well founded ; but, great disciple of Bacon as he was, Davy felt that his opinion was worth nothing unless substantiated by experimental evidence, and, accordingly, he set himself to work to obtain the required proof.

"In Davy's first experiments the two glass tubes which he used to contain the water were connected together by an animal membrane, and he found, on immersing the poles of his battery in their respective tubes, that, besides the now well-known gases, there were really formed muriatic acid in one tube, and a fixed alkali in the other. Davy at once, however, suspected that the acid and the alkali came from a common salt contained in the animal membrane, and he therefore rejected this material and connected the glass tubes by carefully washed cotton fiber, when, on submitting the

water as before to the action of the voltaic current,
and continuing the experiment through a great length
of time, no *muriatic* acid appeared ; but he still found
that the water in the one tube was strongly alkaline,
and in the other strongly acid, although the acid was
chiefly, at least, nitrous acid. A part of the acid
evidently came from the animal membrane, but not
the whole, and the source of the alkali was as obscure
as before.

"Davy then made another guess. He knew that
alkali was used in the manufacture of glass ; and it
occurred to him that the glass of the tubes, decomposed
by the electric current, might be the origin of the
alkali in his experiments. He therefore substituted
for the glass tube, cups of agate, which contain no
alkali, and repeated the experiment, but still the
troublesome acid and alkali appeared. Nevertheless,
he said, it is possible that these products may be
derived from some impurities existing in the agate
cups, or adhering to them ; and so, in order to make
his experiments as refined as possible, he rejected the
agate vessels, and procured two conical cups of pure
gold, but, on repeating the experiments, the acid and
alkali appeared again.

"And now let me ask who is there of us who would
not have concluded at this stage of the inquiry that
the acid and alkali were essential products of the
decomposition of water? But not so with Davy. He
knew perfectly well that all the circumstances of his
experiments had not been tested, and until this had
been done he had no right to draw such a conclusion.
He next turned to the water he was using. It was
distilled water which he supposed to be pure, but still,
he said, it is possible that the impurities of the spring-
water may be carried over to a slight extent by the

steam in the process of distillation, and may therefore exist in my distilled water to a sufficient amount to have caused the difficulty.   Accordingly, he evaporated a quart of this water in a silver dish, and obtained seven-tenths of a grain of dry residue.   He then added this residue to the small amount of water in the gold cones and again repeated the experiment.   The proportion of alkali and acid was sensibly increased.

" You think he has at last found the source of the acid and alkali in the impurities of the water.   So thought Davy, but he was too faithful a disciple of Bacon to leave the legitimate inference unverified.   Accordingly he repeatedly distilled the water from a silver alembic, until it left absolutely no residue on evaporation, and then, with water which he knew to be pure, and contained in vessels of gold from which he knew it could acquire no taint, he still again repeated the already well-tried experiment.   He dipped his test-paper into the vessel connected with the negative pole, and the water was still alkaline.   He dipped his test-paper into the vessel connected with the positive, and the water was still decidedly acid.

"You might well think that Davy would have been discouraged here.   But not in the least.   The path to the great truths which Nature hides, often leads through a far denser and more bewildering forest than this ; but then there is not infrequently a 'blaze' on the tree which points out the way, although it may require a sharp eye in a clear head to see the marks. And Davy was well enough trained to observe a circumstance which showed that he was now on the right path and heading straight for the goal.

"On examining the alkali formed in the last experiment, he found that it was not, as before, a fixed alkali, soda or potash, but the volatile alkali ammonia.

Evidently the fixed alkali came from the impurities of
the water, and when, on repeating the experiment,
with pure water in agate cups or glass tubes, the same
results followed, he felt assured that so much at least
had been established. There was still, however, the
production of the volatile alkali and of nitrous acid to
be accounted for. As these contain only the elements
of air and water, Davy thought that possibly they
might be formed by the combination of hydrogen at
one pole and of oxygen at the other with the nitrogen
of the air, which was necessarily dissolved in the
water. In order, therefore, to eliminate the effect of
the air, he again repeated the experiment under the
receiver of an air-pump from which the atmosphere
had been exhausted, but still the acid and alkali appeared in the two cups.

" Davy, however, was not discouraged by this, for the
' blazes ' on the trees were becoming more numerous,
and he now felt sure that he was fast approaching the
end. He observed that the quantity of acid and alkali
had been greatly diminished by exhausting the air,
and that was all that could be expected, for, as Davy
knew perfectly well, the best air-pumps do not remove
all the air. He therefore, for the last experiment, not
only exhausted the air, but replaced it with pure hydrogen, and then exhausted the hydrogen and refilled the
receiver with the same gas several times in succession,
until he was perfectly sure that the last traces of air
had been, as it were, washed out. In this atmosphere
of pure hydrogen he allowed the battery to act on the
water, and not until the end of twenty-four hours did
he disconnect the apparatus. He then dips his test-
paper into the water connected with the positive pole,
and there is no trace of acid ; he dips it into the water
at the negative pole, and there is no alkali ; and you
may judge with what satisfaction he withdraws those
6

slips of test-paper, whose unaltered surfaces showed that he had been guided at last to the truth, and that his perseverance had been rewarded."

I have at times the mere glimmering of an idea that where a hand-trunk is taken to the bedside of acute illness, filled with an assorted variety of remedies, of cures, whether coarse or refined, there is revealed, in exact proportion to their number, an incapacity to be a Sir Humphrey Davy. Sir Humphrey was a great deal better qualified to study out exact science than remedial therapeutics. The very structure of his mind made it impossible for him to convert inferences into facts; his imaginative powers only seemed to incite him to a more persistent search for facts, for truth. No, he was cut out for a scientist and not a physician, for *science walks by sight and not by faith*—not for him a fusilade at a possible enemy in ambush. Did you ever stop to think what a whirligig route, remedies have to take before they can reach the precise spot of supposed need ?

Take, for instance, a remedy that is supposed to be needed for its local effect upon the small bronchial tubes in the so-called bronchial catarrh. Into the mouth, down into the stomach, there to pause for chemical analysis—then into the blood—into the right chamber of the heart to the lungs ; from the lungs back to the left chamber of the heart, and from thence through the arteries to the diseased tubes ! To expect curative results from such means would involve too much of the imagination, too much loose inference for a man of Sir Humphrey's scientific temper of mind.

Having thus given you a high ideal of the scientific method in the search for truth, and some new conceptions as to the reception and appropriation of truth, and having tried to impress your minds still more deeply of the importance of the subject we are investigating, we are better prepared to go on to a consideration of the fourth parallel.

A gentleman beyond his 60th year was brought to his bed with an obscure disease of his stomach. From his youth he had more or less trouble with his stomach, and for the previous ten years he had been so disabled by his frequent attacks of painful indigestion, as to have become very nearly a mental and physical wreck.

This attack was to be a repetition of several previous ones that had gradually increased in gravity. It had been characterized by pain gradually increasing in severity, until vomiting would occur, when perfect relief would ensue for a time, but to be soon followed in due time by an uneasiness—the low muttering of the distant storm that ended in the usual convulsion. These spells of illness would last generally from a month to six weeks, and at all former ones, feeding had been deemed necessary to support vital power as disabled as was always the stomach. To feed the disabled stomach, to contrive some delicacy that would be forced past the palate in the absence of the sense of taste, or with taste in bristling aversion, was the experience that kept the science and the art of the kitchen in a state of chronic exasperation.

During one attack the Garfield method of support had been tried, with a view to give the stomach that rest it needed, but after a half month's trial it became unendurable, and as there was a general scare over the possibility of starvation (the patient seeming to be wasting away the while), it was given up.

I found my patient wanting either *relief or death*, and as there was a possible ulcer near the outlet or just without it, there was need to prevent the tearing of the diseased structures by the convulsive action of the muscles in vomiting. And then there was defective heart-action; it was going on in a go-as-you-please way, that is, better described by that uncouth word, "wobble."

What was I to do, shut out of the stomach as I was, and the patient exceedingly thin already, from months of deficient digestion? It was my first acute case in that family, and I might find it difficult to defend the stomach from the invasions from the kitchen.

I delivered an address about as follows : "Madam, it will be necessary to relieve the agony in this case, and it can only be done with a hypodermic needle ; it will have the effect to make him really comfortable for hours, and not only this, it will be actually helpful to Nature in a curative sense, because as these spells of vomiting only come on after he has suffered pain for a long time, we may be able to prevent this process by making pain impossible ; in short, we will try to keep the ulcer or the disease, whatever it is, still, while Nature goes on with the cure, just in the same sense that we must keep a wound or a fracture quiet while the cure goes on. As for feeding the patient, your former experiences were a series of annoying disappointments ; you did not nourish him in the least, as the thinning process went right on until natural hunger came, regardless of your soul-crushing efforts to prevent it.

"He will not starve to death for many weeks, and if he dies before he becomes a skeleton, it will be from disease only. This treatment will not only be relieving, curative, but it will save you from the annoyances of the kitchen ; and as we may expect him to be kept comfortable you will care for him with an ease that you have never before experienced."

(Note.—This speech was not delivered until several days of dosing had demonstrated the theory.)

But what was I to do with the heart ? What could I do ? I could not administer digitalis, for the stomach was too sick ; and were I to try to use heart remedies with the needle, my patient, failing to realize any bene-fit, would soon object to being made into a needle-cushion on so large a scale ; and so with very little concern on its account I simply let it wobble on ; and it was not so very long before it toned itself down to a very orderly heart.

The dosing went on day after day, and for nearly a month any attempt to cut it down was resented by the stomach by aggravation of symptoms. The thinning process also went on, but there was little if any appar-ent loss of strength, while the mental condition, though affected by the dosage, seemed to suffer no loss.

At first there was an odor in the breath that could be perceived some feet from the bed. This gradually declined and was scarcely perceptible at the 30th day. After this time the need of dosing rapidly declined, so that after the 35th day it was abandoned. At the *41st day Nature changed her bill of fare,* and there was no trouble in the kitchen to meet the need ; and, I am able to tell you that on this day, there was actual ability to walk unsupported across the room, and that the skeleton condition had not been nearly reached ; while, mentally, he was every way brighter than dur-ing the earlier days of the attack.

Could this possibly have been, had the brain been starved ? Is there any scientific evidence in a case like this that should incite thinkers to thinking ?

Since I began upon the preparation of this lecture, there died a lady physician in my city, of wonderful

acuteness of intellect, from some slow wasting disease.
I frequently saw her in this last illness, not as her
physician but as her friend, and I never saw any
decline in quickness of perception or in interest in all
her human affairs ; yet on that last day of her life the
muscle and fat tissue of the arms had seemingly
totally disappeared. And the skeleton condition was
hard by. Fifteen minutes before death she asked for
the time, and when the beginning of the end was
reached she knew that it had come. She said she was
dying.

Could this have been with a brain also wasted by
disease ? The little starved boy was mentally clear
to within an hour of his death. Could this have been
with a wasted brain ?

Listeners, readers everywhere, have I gone " daft "
on this particular subject, or is there a mighty physiological fact under process of unfolding ?

In the case of my patient there was a gradual
development all along the lines of wasted tissue, and
it did not cease for many months—not until ten pounds
of solid tissue had been added to the ordinary weight
of more than a score of years. This case occurred some
seven years ago, and, though he has had two or three
light attacks since, he was always able to trace them
to an avoidable cause, and now, in his 70th year, after
an average lifetime of martyrdom, he recently informed me that he considers himself strictly a well man.
And all through a higher obedience to the law of his
God "manifest in the flesh."

One more case in this line. My friend, Prof. H. R.
Barber of the Meadville, Pa., Theological School,
informed me that Calvin Cutter, the very popular
physiologist of years ago, *permitted himself to go
through a course of typhoid fever lasting five weeks,*

without taking any food whatever. I was not informed that that experience incited any thinking through which conclusions might have been drawn, that would have induced others to trust nature, even as he had done.

In this last illness of the lady physician she managed her diet in her own way, and with a clear intellect she was able to know what she wanted; she always claimed that her food was taken with a relish, and always without trouble with the stomach, and with body so enfeebled that not the least exercise could be taken. The also enfeebled stomach, even though it were voiceless, was tasked with three or four feedings per day. But further along the digestive line were the recurring attacks of the chemistry of decomposition. Were these a tax on vital power? Did the insensible stomach ever have time to rest into power, between the meals, was there nutrition of the body by such feeding; was there support to vital power where digestion was so slow that the meals must have been received in layers?

It is possible that if the digestive conditions of disease had been duly apprehended, those rare, brilliant powers might have been preserved for years of enthusiastic work in the lines of her ambition and tastes? But medical literature had failed to incite thought, study, investigation; and her medical friend, with all the medical world against him, was unable to incite thought, study and investigation.

The sick must be fed to support vital power; and her life went out, possibly years before the physiological limit was reached. And yet that mind, while it did not trust nature, even as I dared not trust drugs and enforced feedings, was the first to suggest that my therapeutical conceptions were worthy of the printed page.

I have chanced to hear two farewell addresses by patients of mine.   One was by a woman of great intellectual gifts, who reached the dying hour after months of wasting away with consumption.   There was an address to the parents who were to become childless by her death, and to her son, her only child.   Death was known to her to be very near, and yet she was able to talk with an easy flow of language, and with a clearness and directness of thought that was no less remarkable than thrilling.   Was there a wasted brain in her case ?

And I heard a father's parting words to his only son, who stood sorely in need of kindly admonition, and they were words of choicest wisdom marshaled in telling aptness.   He, too, had become a skeleton.   Was his a starved brain ?

LECTURE VI.

My Friends the Readers :—

You have now become somewhat convinced that if the very sick have any power to digest food, it is very much less than you have always supposed ; and have become wholly convinced I trust that the brain has no need of stomach digestion to save it from waste or to conserve its functional powers, even when all else has yielded to the destructive forces of disease. You are now ready for a consideration of the third question naturally arising in this line of investigation.

Is feeding the sick a tax on vital power ?

This question is so involved with the first and second, that it has already received an incidental consideration, and would seem to be in part answered : but the answer so involves the whole field of disease, and remedial therapeutics, that a library might be written upon it.

The physiological limit of a human life is determined in the germ period of existence to a moment ; hence, "length of days" is a matter of heredity in this physiological sense. Did you ever stop to think of the possibilities of heredity ? Let us see : go back eight generations and there are 256 persons to inherit from. What are the possibilities in the line of disease ? of consumption, cancer, scrofula, lunacy, etc. ? Our relatives, perhaps, not so very far back have died in hovels, in palaces, in infirmaries, behind grated windows, of every kind of inherited and acquired disease—perhaps at one end of a rope—and they have been the contending forces on all battle-fields of history.

Mark Twain, our —th cousin, roamed all over the old world with welling emotions, duly guarded for a gushing forth over the grave of Adam !

He who would boast of pure blood and a noble ancestry has pressing need to keep clearly within the lines of credible history or tradition.

The capital stock of a human life, therefore, is its hereditary or constitutional power, and it can never be enlarged, and the vital question with every life is—shall living be confined to the interest, or shall the principal be cut down to meet the extravagances of the days, of the hours ?

Every human life goes on serenely so long as the physiological balance between the constructive, and destructive forces of the body is maintained, by the digestion and assimilation of food. The first step, then, in every disease, is the first loss of this balance by whatever lowers digestive power.

Behind every attack of pneumonia, of rheumatism, behind all fevers, and all malignant diseases that sweep whole families away is heredity, and a taxing from food decomposition, of unknown time, of unknown gravity. It is the " fittest " that survive, and because of hereditary power of self-protection.

To get down to closer work, you must first clearly understand that the digestion of a meal in the most perfect health is such a tax on vital power as to decidedly enervate both mental and physical energy during the active period of this process. What must it be then when this digestive power is weak through heredity, debilitated in the ailing, and prostrate in the very sick ? It is not denied by physicians that *the most perfect defense against disease, is the rich blood of vigorous, perfect digestion* ; and for the present the defense against the deadly microbe ; while we await the germ-killer that will permit life to go on with more indifference to this blood-making power.

Now a case of pneumonia falls into my hands ; it has been made possible by a loss of defensive power of the blood ; certainly, if the theory is correct, this would be so, *but good blood is the product of good digestion.* Weak digestive power, then, must have been subjected to duties that, by as much, must have been an additional tax on vital power, causing a degree of exhaustion of this power, " nervous prostration," hence paving the way for disease.

No one ever gets the slightest sense of refreshment from a meal, without appetite in time of health. All the more it must be so when vital power is taxed, by the extra labor of handling a meal that must be digested very slowly, through impairment of this power, and yet more slowly when the chemistry of decomposition is incited, as in the case of cholera morbus.

Now, in my case of pneumonia, I have a right to assume that the blood has been below the normal in defensive power, for an unknown period ; that this could not have been had not the blood-making machinery been overtaxed ; that it could not ever have been except more food was taken than could have been handled, without the aid of the chemistry of decomposition ; that my patient has been brought to his bed by overeating for an unknown period. Shall I, then, enforce still more eating when all the powers are prostrate with disease ? And, particularly, as there is likely to

be a degree of pain at times that will indicate the relieving doses that paralyze, as you know.

Do you think the brain needs to have any chances
taken of added taxing, that must be inevitable when
food is taken, after Nature, in the clearest, the most
emphatic language, through aversion, forbids?

It was not so very long ago that two children were
taken with acute bowel troubles, that were called
cholera-infantum. One of them, a delicate child of
eighteen months, who was teething at the same time,
was sick enough to have twenty-five actions, tinged
with blood, before it fell into my hands. The other
was also very sick.

There is no other disease that causes such a melting
away of tissues, and therefore that so indicates the
need that the loss be lessened, as much as possible, by
the digestion and assimilation of food, according to
accepted medical science. This child was fed, and with
all the science and art of predigested and other foods,
and the agony duly relieved by appropriate dosing
as far as it could be done. That the treatment was
kept up to the highest level of accepted science, there
can be no doubt.

In my own case there was a degree of nausea at first,
and it was deemed that when so extensive a portion of
the digestive tract was in a state of active disease,
there was no dividing line where the abnormal met
the normal; that the high pulse and intense thirst,
meant total loss of digestive power all along the line;
and that of all diseases it would be unscientific to feed,
where decomposing food would have to be forced along
a bleeding viaduct; and because inflamed, congested,
bleeding, it needed to be held in repose as a wound or
a fractured bone.

Now, with the utmost disregard of accepted methods in treatment, I permitted the stomach that absolute rest Nature seemed to call for, and all the pure cool water the dry mouth, throat and stomach indicated, and they indicated a great deal as soon as the nausea disappeared, which was early in the attack. The intensely alarmed parents were easily made to believe that food not only could not be digested, but for these reasons would be actually dangerous. The child had swollen gums which always incites a nervous condition, and how often during the time of teething do spasms occur, that are generally referred to digestive trouble as the exciting cause! In spite of the best of care this child had a spasm, and for one night and day hope became nearly extinct. The pulse ran way up to be nearly uncountable. What would likely have been the outcome at this period, if the bleeding way had been worried with green cheese in a state of decomposition? Would the nervous system have been unaffected?

The nervous storm passed over after two teeth had appeared, and from thence on, there was a gradual decline of symptoms, and, with the disappearance of thirst, it became possible, for the mother's sake, to admit beef-tea. Now, for many days, this child was so utterly helpless that it had to be gently raised to receive every cool drink. The 13th day came and there was a cry for bread (a favorite food when well), and when it was brought in sight, the child raised itself from its couch with surprising agility to receive it. The disease had lost its grasp on vital power, and there was still a nourished brain, but Nature had changed the bill of fare. And the other child, much older, that had been ill about as long as mine, had its voice stilled in death at the time mine was calling for bread.

Now I cannot know but that my case would have

recovered with enforced feeding, even though the paralyzing dose would thereby have been necessary. But there was apparently so near an approach to death, that I can but believe that the chances of recovery would have been seriously diminished ; and then do you not see, that the brain was kept duly nourished and, while very much fatty matter had disappeared, there was ample muscle-power left for all the need of a healthy growth back to the normal condition.

Nor can it be known that the other child would have recovered, had my method been carried out, for death may have been inevitable from the first. We can only draw inferences from a preponderance of evidence, but inferences at times are strikingly conclusive.

The lady with the cholera morbus that I spoke of had been ailing for years with digestive trouble, in fact had been under medical treatment for months, whereby she was to have the intractable stomach strengthened into power of vigorous digestion through the agency of remedies ; but a case of cholera morbus was not thereby forestalled. But, by living in stricter regard with the physiological conditions involved in digestion, there have since been three years of comfort in the region of the stomach, and a much happier, because a more healthful, life. In her former life the eating was regulated by the time of day or by the whims of a morbid appetite, and never according to the indications of natural hunger. There were headaches, severe colds, days without any eating, nights of fitful sleep, and always a gloomy impression that there was disease behind all that tended to danger, but that could only be met by some remedy somewhere in some skilled hands. And so the wreckage went on.

Having very nearly convinced you that we need not

feed the sick, wholly convinced you that the brain
needs no food at all, other than as provided for in the
fat we can so easily spare, and in the muscles, etc., in
any kind of sickness, until the skeleton condition is
reached, you are now better prepared to understand
the testing of Nature that I proposed to undertake, as
the result of the new views of her powers that were
revived in the case of fever.

It was to be a line of investigation that, so far as I
was aware, no physician had ever undertaken with the
motives that incited me.   True, there have been physicians that have not fed the sick, physicians that have
not used remedies.   I once read that Thomas Jefferson
had occasion to consult a Parisian physician, and was
surprised that the treatment did not involve drugs,
and that none were used in his practice.

I have recently read in a large work, "How Nature
Cures," by Emmet Densmore, M. D., that the sick do
not need feeding, but no reasons are assigned.

You will remember that in this case of fever, I noticed
that there was no unusual waste of the tissues because
of the enforced fast ; that not only was there a gradual
decline of symptoms in every way, but actually a corresponding increase of vital power, as manifest in an
increase of physical strength and a brightening of the
mind.   *Now if Nature were able to support vital
power unaided in so severe and prostrating disease
as typhoid fever, why should she not be able to do
the same in other, in all other, severe diseases ?*   And
with more ease in diseases of lesser magnitude ?

How can I find out without trying, how is any new
thing ever found out without trying ?   Something starts
the mind by a mere hint, as the falling apple the mind
of Sir Isaac Newton, and forthwith, there is a new line

of thought originated and the law of gravitation is discovered. A hint to the mind of Sir Humphrey, and
the world knows the composition of water. It was to
be an investigation in the line of science purely ; it was
to be in the search for a fact that, if established, would
stand for all time to come, and a fact of immeasurable
importance.

It was, too, to be a line of investigation that every
thinking physician, capable of thrashing out even a
large sheaf for a possible grain of truth, would be more
than glad to have undertaken.

But how are you going to carry on such a line of investigation, with all the world against you in a general
practice, where the need to feed is supposed to be measured by the severity of the disease ? Will the friends
of the sick permit the danger of starvation to be added
to the danger of disease ? You may well and naturally
ask this question. You may well think that the
introduction of a method that involved an apparently
dangerous neglect of support to vital power, would
require either extraordinary reasoning powers on my
part, or that very soon I would be considered too dangerous to be intrusted with even trivial cases.

Let me surprise you. In the first place, I had a very
strong ally in Nature herself. In every severe case
the difficulty of feeding was always measured by the
severity. I had but to prohibit milk to let Nature
practically have her own way so as to fully meet
the scientific demands of the question.

Milk is considered a fit food for the sick, not only
because it can generally be taken without aversion, but
because it is a "typical food." Is it? Yes, for the
calf for a few months of its life, but for the human
body only for that period of life when there is little

sense of taste and no teeth, hence the wisdom in the providence of a solid food in liquid form that can be taken without mastication, and that fully satisfies every want and every need. The typical food of hunger of mature life is the food that hunger calls for at each meal, and the mouth of hunger never calls for milk first at the dinner of a loggers' camp. What would be the result upon yourselves, were you compelled to live on milk, however typical it may be, for a whole month while engaged in your business affairs? Would you not go down under it because of its lack of variety, because it would appeal so little to the sense of taste, and therefore to the pleasure of eating? Why cannot we all go on this simple typical article of food and so abolish our kitchen, banish the maids, and hence on, be able to fill our stomachs to distension between two breaths, and so have time saved by the rapid taking of the meals? And how much less expensive would the cost of living be!

Now milk, when taken by the drink, soon becomes whey and a very tough stringy mass of curds, very green cheese, in fact, every mother knows this. Is so solid a food, then, as green cheese to be taken without mastication, a fit food for the sick? And yet patients are kept upon it, and very commonly in cases of fever, week after week, when in cases of health it would fail to be a typical food, because Nature would sooner or later revolt against it. Readers, I am able to say that I was always able to reason milk out of the bill of fare for my sick, when beef and other meat teas were permitted. So little is their power of support, and so averse to the palates of the severely sick, that the largest freedom of use could be permitted with assurance doubly sure, that not enough could be enforced to add danger to the dangerously sick.

Sir William Roberts, in his elaborate work on Diges-

7

tion and Diet (edition of 1891), says, "In a state of health milk must be dealt with in the stomach and the casein is curdled into solid masses ; these masses have to be broken up and to be more or less dissolved in the gastric juice before they can traverse the pylorus." (Query, do they really traverse the pylorus and get into the duodenum, Sir William, without digestion ?) "In the seriously sick, with an almost paralyzed stomach, milk is not meddled with in that viscus. There is neither pepsin nor acid to curdle it, and it passes as a flowing liquid into the duodenum. Arrived there it encounters the secretion of the still active pancreas, and milk is especially amenable to the action of the pancreatic juice."

With the stomach half paralyzed and utterly destitute of pepsin and acid, how is it possible that there can be a "still active pancreas" in the "seriously sick"?

Where is the exact line between the normal and the abnormal in these cases ? Now, when you eat because you think you must, to keep up your strength in your ordinary health, or in the absence of acute symptoms, and when there is only real indifference to food, you always find it a task. There is no sense of relish, and you always arise from such a meal with a feeling of depression, and a wish that you had not taken it.

On the contrary if there is keen hunger you begin to get a reviving sense of refreshment from the very first mouthful ; this means that the chemical change that takes place when the first mouthful reaches the stomach is instant, and the resulting absorption into the blood also rapid. And so this goes on till enough of digestion has taken place in the stomach, to fully satisfy hunger before the last mouthful, and, as you arise from your meal, you have reached the highest sense of refreshment that a meal ever gives. This is in accord with the physiological fact that muscle-forming foods are digested in the stomach, and makes it all the more improbable, not to say absurd, that, in the debility of sickness, muscle-forming foods can be appropriated, no matter how artificially digested, by an excretory membrane twenty-five or thirty feet from this wonderful stomach !

Now the duodenum is only a little tube twelve inches long, a very small organ to serve as a deputy stomach even when aided by a pancreas active with health. A part of its duty is to receive and carry away, as a viaduct, the indigestible rubbish of all the meals, several hours after they have been taken, and hence this process. is never attended with any sense of re-freshment; thus it may be presumed that it, as well as the pancreas, plays a very minor part in the handling of nitrogenous or muscle-forming food in time of health, and a very minor part in time of disease.

You will thus see that the exclusion of milk was not all a mere whim.

LECTURE VII.

My Friends the Readers :—

In forming an estimate of the value of a method in treatment several things have to be taken into account.

1. That most diseases are self-limited, with a tendency to recover. This statement, found in some of the works on the practice of medicine, ought to be placed on the title pages of such works in *underscored italics;* and in raised letters just below, the table, Nature's "bill of fare for the Sick."

2. That behind every attack of disease is the constitutional power, or the constitutional weakness, due to heredity, and the acquired weaknesses, or diseases due to life-long and more or less grave violations of the laws of life, that very largely determine whether the outcome is to be recovery or death.

These are the unknown, unknowable factors, that are to be met with in every case, and are most potent to the making and unmaking of medical reputations. They are the enemies always in ambush ; we can never know just where they are, in what form they exist, how weak or how mighty their power.

Heredity determines what disease, by reason of the constitutional tendency or by an actual structural weakness, local or general, and violated laws, shall

100

occur. There is an undue exposure upon developed, acquired weaknesses—and their results—a cold light or severe, a pneumonia, or pleurisy, or rheumatic fever, etc., with the constitutional tendency, behind, as the determining force, as to what disease it shall be : and what, in connection with acquired weaknesses or disease, the resisting or the curing power shall be, whether recovery or death.

Of the ultimate processes of disease we know nothing. We may see its tracks upon the body in the diseased structures, in the ulcer, in the abscess, in the various forms of skin diseases, etc., and in those diseased structures, found post-mortem, were the unseen forces, the contributing forces in ambush that made death, from the attack of pneumonia or of fever, inevitable from the first. The ultimate action by which a part becomes diseased, or a diseased part becomes normal, is a matter of self-generated cell-power as in the growth of a plant. *How we are to reach the mysterious force with drugs, so as to actually influence it, restore it if lost, is one of the unsolved problems of medical practice.* The ultimate cause of disease has been through all the ages a matter of merest conjecture, until, in these later modern times, we are finding something tangible in the germ theory of the origin of disease ; already we have the bacillus of consumption, the bacillus of diphtheria, the bacillus of cholera, etc.

3. You are also to take into account the fact that the treatment of disease is in a state of constant evolution. The treatment of a hundred years ago with its bloody lancet, with its blisterings, its sweatings, its denials of "cold water to thirsty souls," is the barbarism of to-day, and the treatment of to-day, much of it, is to be the barbarism of the time to come ; yea, it is barbarous even now. There is no evolution in Nature's method

of cure.   She performed her work in the long-ago in exactly the same mode, in cell action, no matter how enfeebled by the withdrawal of pints of the "elixir of life ; " and her power to achieve victory against numerous odds is often surprisingly great.   Even under the barbarous treatment of the darkest ages, she was able with consummate power to restore most of the sick. Each treatment was orthodox in its time, and he who failed to use it efficiently was deemed disqualified for the grave duties of his profession.

But neglect the lancet in a given case, and death occurs ; veritable malpractice is involved.   In my childhood my boy eyes once saw the dark-blue stream spurting into a bowl from my mother's arm, and for some trivial ailment that did not confine to the bed for two days.   Yet that was a treatment that the science of medicine perhaps scarcely questioned at the time. In the early years of my practice, a young friend heavily loaded with fatty tissue, by reason of frequent dizzy spells, determined that I should bleed him because he had "too much blood," and also because it was "too thick."   As he was impervious to argument, a lancet was procured, and, after due study of the topographical region of the arm, a vein was reached after several attempts, through the thick layer of fatty tissue.   The stream flowed, and though much less in amount was drawn than was taken from my mother's arm, there ensued several days of such general prostration, as to settle that mode of treatment in the negative for all time to come.   The dizziness was worse than ever.

And where are we to-day in the evolutionary progress of treatment?   A few years ago, the entire medical world was aroused to a degree of expectancy unparalleled in all its history over a possible cure for consumption, through a remedy that was to destroy its causative

germ, but never were such great expectations so
suddenly dashed ; and only recently there comes float-
ing across the ocean from Europe the stimulating,
the uplifting announcement, that at last the true
germicide of the bacillus of diphtheria has been discovered, and that the Board of Health of a great
city has become aroused. Do you not see that,
pending the further development of the germ theory
of disease, treatment must be kept almost at a stand-
still ?

Diphtheria is a disease which, under different names,
has been treated during many centuries ; and never in a
single case scientifically, if this new treatment proves to
be the correct one. The discovery of new germs is far
in advance of the discovery of agencies that are to neutralize their destructive, disease-inciting power. A
remedy that shall be able to save whole families of
children from being swept into the grave at one fell
swoop by diphtheria, will deserve a monument to the
discoverer higher than was ever erected for mortal
man. *Great is the mystery of cure !*

Among the nerve-stretching diseases that have invited the most decided, the most crucifying treatments
throughout all the ages is rheumatic fever, or inflam-
matory rheumatism. Medical science is ever and naturally on the search for causative agencies, and in this
disease, at last, a peculiar acid was discovered in the
blood, and hence a treatment alkaline in character, that
should reach the blood and neutralize its power, was
the orthodox, the scientific method for many years.
True, it did not take into account the underlying condition or agency behind this acid, that made its existence possible ; but it was a tangible mark to aim at,
and so long as firing is a necessity, a mark always adds
to the interest, and a symptom is better than nothing
for an explosion when powder must be burned.

Then in recent times came into very general use the salicylates, and the medical journals began to teem with their successes, not supported, however, by the Sir Humphrey method in investigation.

But rheumatic fever continues to go in its old ways deliberately, torturingly, with not a day cut down from its own allotted time, for a return to the normal. And we are still without a specific treatment for rheumatism ; and what is true of this disease, or of diphtheria, is true of all others. If the germ origin of disease is to be supported by a germ treatment that shall be efficient, then to-day we are all at sea as to treatments, as we always have been of that most terrible disease, when occurring in the malignant form, diphtheria.

But if this remedy should prove to be the boon we are searching after, and if all other severe diseases prove to be of germ origin, and the true germicide be discovered, whereby disease may be paralyzed at its very inception, then shall we enter upon the golden age of medicine ; for our means of cure will be curative, direct, simple, preventive. We shall then have no more diphtheria, no more cholera, no more consumption, nor pneumonia, for their development can be immediately arrested at their very threshold.

A very distinguished physiologist went into the columns of the " Scientific American " a few years ago over possibilities of this kind, in very roseate phrase. And not only the golden age of medicine but the golden age of living, when life will go on easily, with no perplexing care over what or what not to do for the sake of health ; what or what not to eat to keep the body well nourished, for no matter what we may do, there will always be the ever-ready germ-killer by which disease is made impossible.

4. And you have to take into very careful account

the fact that a physician is never called until the disease has got such headway as to excite alarm in the patient or friends.

In the case of dropsy I referred to, more than a year of perceptible symptoms had been allowed to go on in due development, without the slightest effort to alleviate, and it was only when swelling of feet, hands and body occurred, and ability to sleep in the lying posture became impossible that aid was summoned ; but then what frenzy of effort, what clinging to even straws, what absurd hopes and expectations from the utmost straw in remedial agency !

The time when that case could have been helped, the time when Nature extended her hand for relief from the overtaxing of her vital force, was years before— *help was called too late.*

I was summoned to a case of a little boy suffering from a sore throat that had excited no alarm because an attack of quinsy in an older member of the family had occurred some months before with recovery, and this was supposed to be the same disease. Fatal delusion ! I found the boy still able to sit in his chair or walk across the room, but beyond the power of swallowing even a drink of water. In an hour he was dead with diphtheria ; and the fell hand was not stayed until four of the flock were swept away. What could I do without a treatment ? For many months the only outlet for the gas of a cesspool was into the living room !

There were eleven children in this family, and in a future lecture I will tell you what was done to prevent those without symptoms from taking the disease. But in these fatal cases there was a malignancy that would seem to defy all human means known at that time. What might have been, had there been available this new germ-destroyer! Ah, "what might have been," are often the saddest of words.

In the case of the young lady with the fever at her own home, there were many months of indifferent appetite, of depressing languor, mental as well as physical, in which Nature was getting ready to sum all up in a case of fever, but which might have been avoided had her hintings been duly heeded. I had to take the case on the wing. The voices of Nature in the inception, in the development of disease are of vital importance far beyond the ordinary conception; and when you have failed to call upon your physician in time, make all the allowance that is his right for time lost that ought to have been his, to use preventive means. For behind every attack of disease, are the weeks, the months, and often the years of preparatory work, and work largely avoidable, unless the germ is proven to be the inciting and the developing cause.

5. And you also ought to take into most thoughtful account the physician's limited power in the cure of disease, that you be able to justly estimate his services, and therefore that you err as little as possible in holding him responsible for miraculous results, when he is without miraculous powers.

What, who cures disease? So grave an accident as a broken bone occurs, very grave when the fractured ends are forced through tissues and skin to the surface. The surgeon is called, the fractured ends are placed in apposition and in line, and securely retained by the splints and bandages that the wounded parts may have the perfect rest that is the supreme need. Then the feeding, the work of the stomach, is next adjusted to the change of living habits, then the repairs are all left to the divine hand of Nature.

So grave an operation as an amputation of a limb

takes place, or so exceedingly grave an operation as the removal of a large tumor from the abdominal cavity; the stitches are taken, the straps and the bandages are applied, and, with due antiseptic precautions, with the diet regulated to the changed life-habits, Nature is intrusted with the rest!

But in disease, simple or grave, there must ever be a bombardment or a fusillade if a cure is to be achieved! And with treatments, with remedies that are the science of to-day only to be the barbarism of to-morrow!

Now, readers, I have placed the table before you which I have called "Nature's bill of fare for the Sick." Admitting that it is Nature's sole reliance for the support of vital power, you are struck by the marvelous provision by which the brain, the seat of all power, can keep itself duly nourished no matter how sick the stomach, no matter how changed the conditions of normal digestion, by the absorption and appropriation of those tissues that can be easily spared, and in exact proportion to the need; never calling for the minutest atom in excess! Is there anything more marvelous than this in all nature?

You easily see, then, if this be a fact in physiological science, that beginning my course of investigation with the very next case of fever, I would have a decided advantage in the outset, of having my care of how vital power was to be sustained entirely eliminated, thus permitting by so much a concentration of attention and care upon disease itself, and so be able to study Nature in disease with an interest I had never known before.

There is one more question that I must needs discuss before you can draw conclusions as to the success or failure of my treatment, or care of the sick, and that will be the effect of strong dosage on the stomach itself. As this will involve a consideration of alcoholics as local irritants, I will consider their use generally in my next lecture.

I will close this lecture by reading an extract from a charming little volume in easy rhyme, " Doctor Syntax in Search of the Picturesque," which contains ideas quite in line with some suggested in my present lecture.

The good doctor was making a tour through England on his grizzle mare for the purpose of gathering materials for a book which he was to print, and so make his fortune. He was to

> " Sketch it here and prose it there,
> And picturesque it everywhere."

As it was written nearly one hundred years ago, you will be surprised at a layman's conception of Nature in disease, and a layman's conceptions of the medical methods of his time, who was also a poet.

While making a sketch from a rock on the edge of a pond he fell in, and with the result portrayed in this extract.

> " It was not Vice that e'er could keep
> Dear Syntax from refreshing sleep;
> For no foul thought, no wicked art,
> In his pure life e'er bore a part.
> Some ailment dire his slumbers broke,
> And, ere the sun arose, he 'woke;
> When such a tremor o'er him pass'd,
> He thought that hour would prove his last.
> His limbs were all besieg'd by pain;
> He now grew hot, then cold again:
> His tongue was parch'd, his lips were dry,
> And, heaving the unbidden sigh,

He rang the bell and call'd for aid,
Then groan'd so loud, th' affrighted maid
Spread the alarm throughout the house ;
When straight the landlord and his spouse
Made all despatch to do their best,
And ease the sufferings of their guest.
'Have you a doctor?' Syntax said ;
'If not, I shortly shall be dead.'
'Oh, yes, a very famous man ;
He'll cure you, sir, if physic can :
I'll fetch him quick ;—a man renown'd
For his great skill the country round.'

"The Landlord soon the doctor brought,
Whose words were grave, whose looks were thought :
By the bedside he took his stand,
And felt the patient's burning hand ;
Then, with a scientific face,
He told the symptoms of the case.
'His frame's assail'd with fev'rish heats ;
His pulse with rapid movement beats ;
And now, I think, 'twould do him good
Were he to lose a little blood :
Some other useful matters, too,
To ease his pain, I have in view.
I'll just step home, and, in a trice,
Will bring the fruits of my advice ;
In the meantime, his thirst assuage
With tea that's made of balm or sage.'
He soon return'd,—his skill applied,
From the vein flow'd the crimson tide,
And, as the folks behind him stand,
He thus declar'd his stern command :
'At nine, these *powders* let him take ;
At ten, this *draught*, the phial shake ;
And you'll remember, at eleven,
Three of these *pills* must then be given ;
This course you'll carefully pursue,
And give, at twelve, the *bolus* too :

If he should wander in a crack
Clap this broad *blister* on his back ;
And, after he has had the blister,
Within an hour apply the *clyster :*
I must be gone ; at three or four,
I shall return with *something more.*'

" Now Syntax and his fev'rish state
Became the subject of debate.
The mistress said she was afraid
No medicine would give him aid ;
For she had heard the screech-owl scream,
And had besides a horrid dream.
Last night, the candle burn'd so blue ;
While from the fire a coffin flew ;
And, as she sleepless lay in bed,
She heard a death-watch at her head.
The maid and ostler too declar'd
That noises strange they both had heard.
' Ay,' cried the Sexton, ' these portend
To the sick man a speedy end ;
And, when that I have drunk my liquor,
I'll e'en go straight and fetch the Vicar.'

" The Vicar came, a worthy man,
And, like a good Samaritan,
Approach'd in haste the stranger's bed,
Where Syntax lay with aching head ;
And, without any fuss or pother,
He offer'd to his rev'rend brother
His purse, his house, and all the care
Which a kind heart would give him there.

" Says Syntax, in a languid voice :
' You make my very soul rejoice ;
For, if within this house I stay,
My flesh will soon be turn'd to clay ;
For the good doctor means to pop
Into my stomach all his shop.

I think, dear Sir, that I could eat,
And physic's but a nauseous treat;
If all that stuff's to be endur'd,
I shall be kill'd in being cur'd.'
'Oh,' said the Vicar, 'never fear;
We'll leave the apparatus here.
Come, quit your bed—I pray you come;
This arm shall bear you to my home,
Where I and my dear mate will find
Med'cine more suited to your mind.'

" Syntax now rose, but feeble stood,
From want of food and loss of blood;
But still he ventur'd to repair
To the good Vicar's house and care;
And found at dinner pretty picking,
In pudding boil'd and roasted chicken.
Again 'twas honest Grizzle's fate
To take her way through churchyard gate:
And, undisturb'd, once more to riot
In the green feast of churchyard diet.
The Vicar was at Oxford bred,
And had much learning in his head;
But, what was far the better part,
He had much goodness in his heart;
He also had a charming wife,
The pride and pleasure of his life;
A loving, kind, and friendly creature,
As blest in virtue as in feature,
Who, without blisters, drugs, or pills,
Her patient cur'd of all his ills.
Three days he stay'd, a welcome guest,
And ate and drank of what was best;
When, on the fourth, in health renew'd,
His anxious journey he pursu'd."

LECTURE VIII.

*" Wine is a mocker, strong drink is raging ; and whosoever is
deceived thereby is not wise."*

ALCOHOLICS CONSIDERED—PROF. N. S. DAVIS QUOTED—AUTHOR'S
VIEWS AS TO THEIR EFFECTS IN DISEASE AND IN SHOCK—STORY
CONTINUED WITH ILLUSTRATIVE CASE.

My Friends the Readers :—

Prof. N. S. Davis of Chicago is easily the Nestor of
the Medical Profession of America.   He is not only an
M. D., but an A. M. and LL. D. as well.   For many
years he was Professor of Practice and Principles of
Medicine and Clinical Medicine in Chicago Medical
College.

He is the father of the American Medical Association, and was made its presiding officer at several of
its meetings.

Because of his eminence in the profession he was
selected to preside over the meetings of the great
International Medical Congress that met in Wash-
ington, D. C., in 1887.

He has written a work on the Practice of Medicine
that is an authority among physicians the world
over.

Among the apostles against the use of alcoholics he
is pre-eminent.

That he has not been assigned to the ranks of
" temperance cranks " years ago, is because of his gi-

gantic proportions in the intellectual life of the profession. He has been too large a man to be so disposed of by smaller men.

In his work on the practice of medicine he writes this golden text (I quote from memory) :—"*A physician may have the highest possible success in the treatment of any kind of a disease without the use of any form of alcoholics.*" My dear doctor, why did you not print this golden text in golden letters on your title page ?

Prof. Davis has been my referee, my rock of support, whenever I have been called upon to stand against the use of alcoholics, and for many years. I have had to defend always, when, in a severe case of sickness, I have had counsel ; for the professional brethren of my school in Meadville, without an exception, believe that alcoholics support vital power ; hence, in the use of alcoholics, my treatment in their estimation has been defective.

Now here is the latest science of the effects of alcoholics, according to Prof. Davis, which I find in his introductory address before the International Congress :

"It has long been one of the noted paradoxes of human action, that the same individual would resort to the use of the same alcoholic drink to warm him in winter, to protect him from the heat in the summer, to strengthen him when weak and weary, and to soothe and cheer him when affected in body or mind. . . ."

"The alcoholic drink does not relieve the individual from cold by increasing his temperature, nor from heat by cooling him, nor from weakness and exhaustion by nourishing his tissues, nor yet from affliction by increasing his nerve force, but simply by diminishing

8

the sensibility of the brain and nerves, and thereby lessening his impressions of all kinds, whether from heat or cold, weariness or pain. In other words, the alcohol by its presence does not in any way lessen the effect of the evils to which he is exposed, but directly diminishes his consciousness of their existence, and thereby impairs his judgment concerning the degree of their effect upon him."

Words more alive with the white heat of truth, were never penned by mortal man.

There is one statement in this quotation that I wish you to plant in your minds where it will take root, namely : that these effects are produced by *diminishing* the sensibility of the *brain and nerves*, hence alcoholics are in a sense anæsthetics and not stimulants, that they paralyze, and do *not support brain power*.

No human being ever takes a drink of an alcoholic with an idea that his need is to be met in a curative sense, for he knows by bitter experience that it is not medicine for him. He always takes it, that his afflicted sensibilities, whether of body or mind, shall be for the time "drowned," knowing all the while that they never die under water ; rather, he takes it that they shall become so benumbed that life can be better endured the while.

Now I want to call your attention to the local effects of alcoholics upon the stomach.

They act always as irritants to its delicate mucuous lining. The irritation causes a congestion, a swelling of the capillaries that are so small and finely meshed, that the sharpest needle cannot enter the membrane without wounding one of them. Their walls are elastic, and hence irritation or pain causes such an active determination of blood that they become at once distended. Now, small as they are, there are still smaller tissues between them, which at once become subjected to a strangling pressure. The gastric glands also become subject to this pressure, hence the sources of their supplies whereof the solvent juices are generated, are diminished, and perhaps often entirely suppressed. This is an incipient stage of disease, and in proportion to the frequency, the power and the continuance of the dosage, so must be added an abnormal length to the course of disease, for there must be a normal stomach behind normal health, and it must get well, before its work can be done with power.

But this is not the new point against the use of alcoholics that I promised you, but you have not probably seen this fact so stated before, and this applies equally to all medicines that can be disagreeably felt when they reach the stomach.

The stomach, my readers, is a mighty power in convalescence, when its delicate mucous membrane has been preserved with physiological care. Now, listeners, you are going to admit, for the sake of my argument, that the brain does not need any support through the digestion of food by the stomachs of the sick, and since alcoholics lack the indispensable nitrogen to make them in the least nutrients, and since they are never given as medicines in the curative sense ; that they are never given except as support to vital power ; and as, according to Prof. Davis, they actually paralyze instead of adding support to it, how can I, how can any one, logically, physiologically, ever prescribe them ?

If they are stimulants as generally believed, would they do more than to excite the machinery to a more rapid and needless waste of the brain food ? If a depressant to brain power, can any dose be other than harmful ? Every dose taken in the time of normal health, has its depressing effect on vital function, a fact no one will care to dispute. Why then not all the more grave when the resisting power to the effect has been lowered by disease ?

Alcoholics are anæsthetics.—The brain will care for itself until the body is nearly absorbed, when disease does not paralyze its power. Must I then give these paralyzers of brain power in diseases that are very generally self-limited, that are very likely to recover, when such doses cannot be administered without structural injury to the stomach, and are exhaustive to brain force ? "Tell it not," do not even whisper it, "in Gath." "Publish it not in the streets of Askelon."

The citadel of the demon Alcohol has always been, the supposed need in the shock of injury and in the crisis of disease, the tiding-over, the bridging-over period of Nature's supposed extremity. The flagging heart must be kept in motion. It is only in modern times that the guns of the scientists in medicine have been turned upon this citadel.

Says Prof. Davis in this same address: "Speaking of the very generally accepted doctrine that alcohol is a cardiac (heart) tonic, capable of increasing the force and efficiency of the circulation, and therefore of great value in the treatment of the lower grades of fevers, the more recent experiments of Professors Martin, Sydney, Ringer, Sainsbury, Reichert, H. C. Wood and others, have clearly demonstrated that the presence of alcohol in the blood as certainly diminishes the sensibility of the vaso-motor and cardiac nerves, in proportion to its quantity, until the heart stops, paralyzed, as that two and two make four ! !" And that "nothing could more clearly demonstrate the power of alcohol to paralyze both the respiratory and cardiac organs (or the lungs and heart), than the experiments detailed by Dr. H. C. Wood in his address to the recent International Medical Congress at Berlin, on the subject of anæsthesia."

Again, "After an ample clinical field of observation in both hospital and private practice, for more than fifty years, and a continuous study of our medical literature, I am prepared to maintain the position that the ratio of mortality from all the acute general diseases has increased in direct proportion to the quantity of alcoholic remedies administered during their treatment ! !"

And that "both the popular and professional belief in the efficacy of alcoholic liquids for relieving exhaustion, faintness and shock, etc., is equally fallacious ; all these conditions are rapidly recovered from by taking the recumbent position and having free access to fresh air. Ninety and nine out of every hundred of such cases pass the crisis before the attendants have time to apply the remedies. . . . Indeed, whenever a person affected by sudden exhaustion or syncope, is able to swallow wine or whisky, he is in no immediate danger of dying, and yet the recovery is always attributed to the last remedy given, even though its real influence may have been injurious to the patient."

The only evidence I have read, or ever heard spoken in support of the use of alcoholics that has even a hint of reason behind it, is that in certain diseases where the heart seems to be losing power, as in low fevers, etc., the effect of an alcoholic is to lower the pulse, increase its volume, and that this is the evidence of its need. I have never had occasion to try this treatment myself, but, if such is the fact, then the use of a remedy in disease that in health will throw muscle, mental and moral power into general confusion, as revealed in every case of drunkenness, must involve some anomalous power of action, if curative in its effects.

The second point I have to offer against the use of alcoholics is their supposed need in these cases of shock, as from injury, and in the crisis of disease which the professor says will be passed, and begin to revive before there is even time to try to relieve.

(My dear doctor, ninety and five per cent. of the medical profession are against you in this statement— how about a preponderance of numbers in a matter of evidence? Perhaps you will reply that the ninety-five per cent. hold conclusions not drawn up on the Sir Humphrey Davy method.)

A human life begins with a single cell, that, by virtue of its self-generative, self-sustaining, living force, increases by the multiplication of division, each part only developing for further division until a human "form divine" is reached.

The destructive, constructive change that goes ceaselessly on in cell-life incites a need whereby we must breathe; hence self-generated cell-power is the vital force behind lung and heart action. In cases of shock from injury this force is more or less paralyzed, and the degree determines the question of life or death. According to the degree, cell action will so go on as to determine how much breathing is to be done to meet the demands of oxygen. Now, please listen. The crisis dose is always given to keep the flagging heart in action, but the lungs will act if there is a call for oxygen, and with convulsive effort if there should be forcible restraint; if the lungs act, the heart must. There will be no call for oxygen unless self-generated action of the cell creates it. Why then give alcoholics and on what power does it act? *Only on power already in action, that is certain to get back out of the crisis period where death is not inevitable.*

Grant that alcoholics are stimulants in the sense ordinarily supposed, why needed if they can generate no cell power, and if they act only on power already in action, but so enfeebled as to make rest the imperious need?

Shall we apply the whip to the horse ready to drop from exhaustion? Have the heart and lungs need of such help in times of crisis?

I have seen a patient breathe hour after hour with measured movement, who had endured crisis dosing by the pint, until, there was no longer ability to swallow, yet still live on when dying seemed to be only a sleep of health except when aroused by the prick of the hypodermic needle; and still live on hour after hour when the hypodermic barrel of whisky had become absurd, dead everywhere except in the ultimate cell! The heart beating, the lungs expanding and collapsing in exact accord with the demand of oxygen as incited by cell change, even as in health! And I have seen such cases often. Does this mysterious power behind lung and heart action ever need then the paralyzing dose of the alcoholic? Can it ever be administered in the shock of injury or in the crisis of disease, without adding an additional tax, an additional danger to the already partially paralyzed self-generating force of the cell?

It is my view then that alcoholics have no power over cell force, to determine whether life or death is to be the outcome.

The conclusions in support of the supposed need of alcoholics in cases of shock and in the crisis of disease have not been drawn by the Sir Humphrey Davy method.

Now, readers, you are still somewhat better prepared to form an estimate of the success or failure of my new method in the treatment of the sick. You must see that in all cases where the mucous membrane of the stomach was not structurally injured by the irritation of alcoholics, and of strong medicines which acted also as irritants, I must have gained not a little in the convalescing or growing back to the normal condition. Of this I have no doubt myself—it could not have been otherwise.

Now it so happened that my very next case of importance was one of fever, a lighter malaria. The patient, a young married man, was quite content to trust Nature, with my assurance of its safety, when at the same time his aversion to food was strong. And so the days went on with only light dosing, but with the purest water that was available, the mental and vital strength increasing with the decline of thirst until, at about the end of the third week, Nature changed her bill of fare.

As case after case revealed the same result, and always with a more rapid decline of severe symptoms than was experienced under the old method of feeding and heavier dosing, I began to realize that the new method was far more effective in lessening the force and duration of disease, unless I were getting a change of type of disease to deal with ; but certainly I met no change in the types in the onset ; it was only in the more rapid decline and a lessening of the duration that indicated an improvement. Even severe cases of rheumatism seemed to let down a little in severity, and often fixed duration. Fully believing that the supposed morbific cause was only one of a series of symptoms that had behind it an unknown primary cause, I utterly ignored all specific medication and kept my patients comfortable, by the relieving dose as indicated by agony.

I can have no doubt that feeding is a disease-prolonging agency, hence, if a fact, then my cases all must have been shortened by as much as this tax was avoided.

Rheumatism, I believed, was due to disordered digestion, or arising from eating for an unknown period beyond the power to digest and assimilate, hence a disease, determined in character by constitutional tendency, hence the need to obey when Nature called a halt, in causation, until repairs could be made. And why follow the old ways of treatment? Have they shortened by an hour the days and nights of agony during all the history of treatments?

Then do you not see there was the possibility of the germ origin and the germ cure, and, pending this, why not trust Nature to do something of the curing herself, even as we trust her in the fracture and the amputation?

One thing is absolutely beyond any question, that from the time when the loss of tissues disappeared by the cessation of disease, and therefore the return of that appetite that called aloud for the food which is best worth the digestive process, from that time until the absorbed tissues were duly replaced by those as fresh, as instinct with life as if just formed and fashioned by a divine hand; the time was not only shorter than the average, but living itself became a luxury of life, an existence of child-like happiness and indifference to all the strain of human affairs, in every case.

I had one opportunity to study this disease in a com-

parative way.   I was called to attend a case in a young
woman suffering a third attack—one in childhood, next
in mature age, and this the severest in onset of all.   In
the previous attack, which lasted two months, milk
was freely given, day after day, during all the long
weeks.

In this third attack a debate was needed in which
clear definition and unanswerable force of statement
and reasoning was necessary, as to whether my patient's stomach should or should not be converted into a
factory for the making of cheese curds, *not marketable*,
for unless milk be given, how was life to be maintained
for another 6 days?   I won the victory for Nature,
and then for a number of days, when there would have
been agony far beyond what was experienced before,
but for the dosage that made even the sick-bed comfortable, Nature went on with her work, there being no extra tax of vital power in the region of the
stomach.   And what the outcome?   In less than a
month the dosing was easily suspended, and soon there
came, even before the end of the fourth week, an appetite for such food as rejoices the mouth of the wood-
chopper, and the doctor bowed himself out.

During the five or six years since, there has been no
hint of a recurrence of the disease, for there has been
a closer walk with those laws of life whose infringe-
ment made the disease possible.

During the past 17 years many cases of rheumatic
fever have been treated on this plan of non-alimenta-
tion, and with the most satisfactory results, having
found that, by not taxing vital power with food, and
by the avoidance of those special remedies for the disease that play havoc with the stomach, not only has
there been a shortening of the symptom stage of the
disease, but by a good deal the convalescing stage.

And in these later years by teaching a higher science in prevention, through an unfolding of the physiological mode in cure, recurrences of disease, no matter what the form or character, have been rare to what they were formerly. The sick-bed ought always to be a text for a dissertation on cause, as well as cure, and thereby of avoidable recurrence.

LECTURE IX.

My Friends the Readers :—

As the months and the years rolled on, and case after case of severe sickness passed through to a more successful issue, than I had ever experienced before, so did my faith and wonder at Nature's power in disease enlarge ; and so did my love increase for the practice of natural science in disease, because I could but believe I was vastly more helpful to Nature than before ; and so did it become more and more clearly evident that, for some unfathomable reason, Nature had no need of my artificial attempts to support her vital machinery when a contest with disease is going on ; and so did it become more and more clearly evident that, as symptoms declined, so did power increase all along the line, even in some cases, for weeks, unaided by the digestion of food ; and so, anomalous though it appeared, it began to be very evident that a new fact in medical, no, natural, physiological science had dawned upon me.

And how could this be ? I utterly failed to satisfy myself ; indeed, the mystery only seemed to deepen as it became more evident as a fact in science, and it was not until my eyes first glanced at the table I have told you so much about, that an explanation was possible. and *that* made it clear as an object in sunlight, at least in my mind.

124

And now, after a study over the question for years, including a study of all my fatal cases, it is my strong conviction that death is always due to the paralyzing effect of disease or injury on vital power, in a sense that opium, chloral, etc., paralyzes, and not because the brain is starved ; in other words vital power becomes so overpowered that the brain becomes disabled from withdrawing its supplies from the several tissues by some unknowable process of cell destruction or paralysis.

And this reminds me that I must speak of my fatalities, for success of a method in practice is always gauged, you know, by the mortality diminished. " Nothing succeeds like success."

During the past seventeen years my fatal cases have been mainly confined to infancy and childhood, to such diseases as diphtheria, convulsions, croupal bronchitis, etc., with illnesses generally short, and to such diseases as consumption, cancer, etc., in mature life, and from old age.   Now, when a case of sickness falls into my hands and the patient dies before the skeleton condition is reached, it will not be from starvation : and when, after all taxing of vital power through the alimentary canal has ceased, and I add no more to it, and the patient is kept under the best conditions of care, with pain duly kept within bounds, then, if Nature fails to cure the disease, just as she does the wound or the fracture, it will be because death was inevitable through hereditary or constitutional conditions, and acquired conditions, avoidable or otherwise, and never because I have failed through lack of enforced feeding, or through failure to bombard the symptoms.

In my acute fatal cases, there was so general an inability to feed, even if it were desired, that I cannot now recall more than two or three cases where a shortage in sense suggested that I had failed through lack of feeding—but how very frequent it is with all physicians to be called to account for failure somewhere in the line of treatment.

During my studies of Nature in disease, I was not without occasional opportunities to draw inferences by way of comparison.

I have frequently had the necessity of calling counsel thrust upon me, and always, especially in my cases of fever, the lack of milk and alcoholics was found to be an error on my part that had to be corrected ; but in no severely sick case was the patient able to endure the change of treatment.   In one case of fever, the most severely sick that I ever saw recover, counsel ordered milk freely, and whisky even to a pint in twenty-four hours, if it could be borne, but both had to be speedily abandoned ; and though thereafter vital power got so low that there was no conscious sense, the eyes remaining partly open hour after hour, nearly glazed, and death seemed inevitable, behind all there was an unknown, unknowable force not exhausted, a constitutional power that determined that recovery was possible ; and, readers, when all the symptoms were relieved, when there was a clean tongue and no more thirst, and there was the first keen relish for the best of food, there was a degree of physical and mental power achieved, infinite, almost, as compared with the lowest state when even a breath might stop the breathing.

Then occasionally I got within range of a case where some loose inferences might be crudely drawn by way of comparison.   I was called to a case of typhoid fever in a young girl.   Hard by was a case that had been under treatment for a week with a stomach that was still ejecting curds and whisky. As the case proved to be a protracted one, not only in this but in all other respects, I have no reason to doubt that the treatment was well up to the latest science of therapeutics. Vital power "must be supported," and milk was the "typical food;" alcoholics, of course a support, a necessary reinforcement.

I put no trouble into my patient's stomach and therefore no trouble came out of it. The fever was severe enough to cause mental aberration for a week or more; there was a high pulse, a high temperature and the typhoid tenderness. And also there was the decline, in due time, of all the symptoms, and *near the close of the fourth week a return of appetite announced the cure achieved.* There was no food given, and so little of remedies that the cure was fully assigned to Nature, and just in time to relieve the mother's care, to undergo, herself, a run of the same disease. She was about thirty years of age and heavily loaded with fat, and hence a bad subject for a fever; and especially after the days and nights of broken rest and sleep, and of depressing anxiety. She came to her bed with great general exhaustion of vital power, and her case proved very severe, so that during the second week there seemed little reason to hope.

But as in the other case no food was given, but water, water by the quart, by the gallon even, as Nature demanded. But there was the unknown constitutional power behind the disease, and *during the fifth week Nature changed her bill of fare and a cure was achieved.*

What of the other case? A few days later death closed the scene. Now this was an older patient, and death may have been inevitable from the first; there may have been the unknown contributing diseased conditions that only a post-mortem would have revealed, as I so often found in the army ; but with every possible allowance made in this way, the comparison to be drawn from the nature, the science, and the art, in these cases is at least suggestive.   Could I possibly have aided Nature to more hasty work by the supporting treatment used in the fatal case ?

The more I study the question of nutrition in disease at the bedside of acute illness the more am I unable to comprehend the logic of giving the sick, and especially the very sick, a form of food that even in the most vigorous health cannot be borne, even for a single day, without a lowering of vital power ; nay, that where even one meal of it cannot be put into the stomach of hunger without a clearly perceptible loss of power.

In health we need our bill of fare changed somewhat at every meal to have it " typical," so as to meet the demands of the relishing sense.   Yet here is a form of food that is to be taken day after day, week after week, and with aversion or indifference, to be enforced even as a medicine ; too irritative, too taxing for the stomachs of the well, and yet a fit food when the digestive function is prostrate with disease.

No physician will admit that normal health can be maintained for a single day, for the above reasons, on milk and whisky ; then where is the logic of feeding it to the sick ?   How expect, by its use, to raise abnormal health to the normal, when it inevitably lowers the normal to the abnormal ?

My own conceptions have got to undergo a great revolution before I can see in milk and whisky a fit form of food for a human being, sick or well.

I was once called upon to advise in a case where a young man was in bed after a six weeks' illness of so-called malarial fever.   The physician was also a young

man and this was one of his first cases of severe illness.
I afterwards found that the stomach had endured a
bill of brandy, whisky, bromides, quinine, opiates, chloral, etc., to the extent of very many dollars, so many
that the druggist might much better rejoice than the
luckless stomach.

I found that ability to feed longer was very nearly
exhausted, and I persuaded the young physician
that his patient's stomach was so very much like the
condition of the grass of the schoolyard during term-
time, that he became willing to order a vacation ; and
all the more as his patient was evidently failing after
six weeks of authorized treatment, comprehensive in
its scope and forcible in its execution.

*It only required three days of rest which, strange to
say, was occupied by the patient in adding somewhat
to his general strength,* when came a demand for food,
and from thence on, as the disorganized stomach grew
back into power, so did the lost tissues become restored.
But it required time to undo the needless wreckage.

In this case there had been a degree of nervousness
that called into use chloral and opiates to still, and which
was wholly due to the avoidable taking of milk that
could not be digested, the brandy, opiates, etc.   When
the stomach is kept empty it can be entirely ignored if
there be need to use remedies to relieve pain or to enforce
sleep, and this need is greatly lessened if there is no
needless exhaustion of power going on in the stomach.
But with the stomach kept empty the brain will care
for its own nourishment, even when under the influence
of the pain-relieving dose.

You are now prepared for the suggestion that when
you are taken with a cold that sets you to undue and
annoying coughing, you should never put your cough
doses or pellets into a busy stomach ; and you are to

9

pay no attention to the adage, venerable with age, that you are to "stuff a cold and starve a fever," for a classical scholar has informed me it has got into English in a dislocated sense, the real meaning being that if you stuff a cold you thereby have to starve in a case of fever.

I will now tell you about a case where my faith in Nature was to have an opportunity for a crucial test.   Heretofore all my studies had been over cases where the chances were borne by other hearts; now mine was to be put to the test.

Diphtheria invaded my immediate neighborhood, and there was mourning because idols were swept away.   My son of five years was seized upon.

Now this son, by reason of a sunny temperament and a most happy originality in always having his own way without ruffling other ways than his own, had linked his life, his soul with my own in such continuous, strong and tender union that I could not remember up to the time of this attack that he had ever given me anything but sunshine, nor that he had ever received from me anything but caressing.   And now he was in the toils of this disease without a remedy. "No remedy do you say," why even to-day, October 23d, 1894, all Philadelphia is in alarm over the possibility that diphtheria may get permanently located in the city, and has become excited over the news of a new germ destroyer that comes wafted from across the sea.

Ah, Gentlemen of the Board of Health, send over across the waters your committee of investigation, but organize yourselves into a home committee to investigate avoidable cause.   Diphtheria attacks the enfeebled, and never the vigorous.   Rich blood is your true defensive armor against the deadly microbe.

Rich blood awaits upon vigorous digestion; vigorous digestion only comes from a conformance with the physics, the chemistry and the mental forces involved, and this is possible to a degree far beyond the professional conception, very far beyond the general conception. Only the strong man is armed and the "fittest" survive because of strength.

You will never stamp out diphtheria with a specific, and the malignant cases will die because hereditary and acquired conditions have made death inevitable.

In this hour of my trial two medical friends called who had all the ability, the learning and the great experience that could add efficiency to their services and honor to their calling. And they gave me all that the very latest accepted medical science had to offer; and coming down through the ages, the accumulation of centuries of experience in the trial of treatments, science could only offer me quinine and iron in solution, whisky and milk to support vital power! That was the sentence. Each dose would require a holding of hands, a prying open of the mouth, perhaps a closing of the nostrils in order that it might be forced down an ulcerated, a bleeding way.

My friends left me, and with this human life in question, *to stand alone with the medical universe against me.* There was the victim and but one person to stand between him and the barbarism of accepted treatment for diphtheria. Did my faith in Nature fail me? No, far from it. Never did I so fall back on her for uplifting support; it was my only comfort that I who had never given the sufferer anything but kindness in health, was now to be his strong arm of defense from cruelty when he needed kindness as never before.

Let us look into the matter of the science of dosage in such a case. There was an ulcerated throat. In case of an external ulcer the science of cure requires the most perfect freedom from violence. In this case science was to be reversed, and doses were to be given that would set in motion muscles beneath the raw surface, with nerves on fire, that would lacerate in all directions; and this to be done every one or two hours day and night! And there was the quinine, anabomination of bitterness, and not an *emergency* remedy. It can be given day after day, and the good effects have to be seen with the eye of faith, aided by a powerful imagination.

And the tincture of iron that has to be piped through the mouth to save the teeth! What was it to do, what could a stomach paralyzed with diphtheria do with any drug anyway?

Now it so happened that I left the house without taking the prescriptions along, expecting my lighter dosing to go on, and, not returning as soon as expected, the anxious mother had gotten them filled and one dose down, and a tempest raised that took more than sixty minutes to calm. Since there was no germ killer any more than there is now, unless a new light has dawned in ways that are dark; would the *support* of such a dose be likely to be balanced by the *expenditure* of the giving and that would justify me in continuing? That was the end of the science of this case. I realized the possibilities, and if it were to be mine to have to take the final look into a casket, there was to be no memory of manacled wrists, of appealing, wondering, reproachful looks because of cruelty not understood.

I had no remedies, and the unmedicated doses thence on were to satisfy the mother's sense of need. "But did you use no local germ treatment?" you ask.

In matters of external decomposition there are the

crows, the buzzards, and the lesser and lesser scavengers, down to the microbes, all to enjoy the same feast. Has Nature not provided for internal decomposition? There are the microbes, the non-infecting, that are the natural habitats of the mouth. Do not these increase because of a need? Is it not possible that the real essential cause is too subtle for the microscope, and that, after all, the infecting twin-brothers of the native inhabitants may also act in the same sanitary sense? There was no annoying fruitless local treatment.

My son was a very sick boy, there were several post-nasal hemorrhages, there was the fœtid odor and the characteristic depression of vital power, *but my sheet-anchor, Nature, was all-sufficient, and a life was saved.*

My last experience with this disease was in the family already mentioned. When the fourth member was attacked another physician was promptly called who had sole charge, while I kept the second and third cases in my charge. His was immediately put on the milk-whisky treatment, which was borne only a few days, for dosing had to be given up, because of inability to swallow when it had to be without treatment several days before death occurred because none could be given.

In one of my cases, a young lady of twenty, ability to swallow was lost during the first week, and from that time life went on for two weeks without even ability to get one swallow of water into the stomach.

Some of our authors have persuaded themselves that enormous amounts of alcoholics have saved life in some instances, because they were borne without the usual effects, therefore they must have met an indication of the disease. Perhaps there was impaired

absorption power.   Perhaps the defensive power
against all attacks was able to win victory ; such proof
at best is inferential.

The most serious, the most desperate-appearing
cases, often recover, even when no treatment could be
borne.   Alcoholics, within the sphere of my knowledge,
have been generally used to the level of the capacity
of retention.   They have won no reputation as
specifics.   Those cases that can bear the enormous
dosage are so exceptional that they form too poor a
basis for even inferential evidence.

LECTURE X.

My Friends the Readers :—

I have now given you all the theory, the philosophy, the argumentative and illustrative evidence I shall take time to offer on the subject of Nutrition in Acute Disease. *Behind all are seventeen years of attendance at the bedside of the acutely sick*, during which the number and variety of cases cared for were most ample for the questions involved.

During these years, there was not lost more than one week per year from this service by vacations, and in not a single instance was a case neglected through any disability of my own—and this statement will apply to the eleven previous years. And yet more, the practice of medicine, or to express the term in a different way, attendance upon the sick as a physician, has been the sole business whereof I have earned my bread of life, unless a service of six years as secretary of a board of U. S. Examining Surgeons for pensions was a diverting, additional service.

These seventeen years were years of study, of thought, of sharp, keenly interested observation ; of a clearing away of the confusion, of the fogs and the mists of doubt as to the mystery of cure. They were years of such study as every thinking, high-minded physician would rejoice to have go on, for the line of investigation was original, and the object, the ad-

135

vancement, and the enlargement of the science involved in attendance upon the sick.

You easily saw that after the illustration given me of Nature's ability to keep her machinery in motion while engaged in a contest with one of the severest of fevers, and also of her power to cure without the ordinary dosage, I could enter upon this course of investigation supported with a good deal of reason. Indeed I actually entered upon it with very little doubt as to the involvement of any hazard to my patients.

I have given you a few striking illustrations of this power where I have had many—in fact where I had all, every case, as an illustration, except the fatal ones. And let me assure you, in measured words alive with conviction, that that long series of cases running through seventeen years of attendance has been a line of evidence, line upon line, of the self-sufficiency of Nature to right herself in attacks of disease, no matter what the disease, or how severe its character, when constitutional or acquired conditions, against which treatments would be powerless to avail, did not make death inevitable, and without feeding or dosage.

That is high ground and strong language—yet there was no fatal case during these years that could possibly be tortured into evidence against it. Nor was there a case that, by undue prolongation, could be arrayed as evidence against it.

If results so claimed had only the support of treatments, you might well discount them. In all this you have been invited to look upon Nature as I have tried to reveal her, as a finely lined object in perspective under the best conditions of light.

I have tried to open your eyes to the enormity of reckless, meaningless medication, by those illustrations where success was achieved, in which medication was impossible, and also by the suggestion that science is scarcely to be found in treatments that are scientific ; authorized to-day, but are obsolete, barbarous to-morrow.

Some of you could but feel that I got beyond reason, that there was almost criminal rejection of treatments in the case of my own son. You fairly trembled for me, but now what ? Is not the entire medical world aroused over a hope that the new method in diphtheria is to be to this fell destroyer of homes what the vaccination of Jenner has been to small-pox ? Fondly do we hope that the absurd expectations of the Koch method in consumption is not to be the outcome—as we are to-day without any remedy or treatment for this disease, that is in a shade specific, unless this new one proves to be the long-looked-for one. Was I other than an armed guard, standing alone, beside the couch of my prostrate, bleeding, and, for aught I knew, dying son ? I felt so then, and all the years of experience since has only added to the impression.

I have tried to make you clearly see by comparative illustration, *that in proportion to the intensity of disease so is digestive power diminished ;* and that feeding the sick under those conditions is by as much a tax on vital power that must hinder, and thereby prolong Nature's efforts in the cure, and inevitably add undue length to the course of the disease.

And not only this ; your attention was called to the fact that *feeding never prevents the melting away of the tissues under these conditions.* And finally you were shown that *wonderful bill of fare* that furnishes the true source of vital power, that fully explains the secret of Nature's aversion to all eating in her times of battle for life against disease.

Calling upon you to admit this as a fact in physiological science, I have tried to make you understand by
as much that alcoholics are not, cannot possibly be,
other than depressing forces to Nature in the active
stage of disease; and that, by their injury to the
stomach membrane, disable it for that vigorous work
that it should do when disease has lost its grapple
upon vital power.

As a reinforcement to this inference, I have given
you, in condensed expression, the latest views of the
master scientists in medicine, that alcoholics are essentially paralyzers of vital power, that they are not
stimulants but anæsthetics whose real effects are to
benumb all disagreeable, mental or physical sensa-
tions, and to cause by as much inco-ordination of all
the mental, moral and physical forces, as every case of
drunkenness so strikingly illustrates.   Hence you
were called upon to believe that, even in cases of shock
from injury and in the crises of disease, their use
must be attended with danger in proportion to their
severity.

And I tried to take you back to the very source of
power, the self-generated power of the cell itself, and
make you see that this power has no need, can have
no need, of an alcoholic to determine whether it shall
go on or how go on ; that life or death must be determined in all cases by the degree in which injury has
affected it, and never by subjecting it to the force of
an anæsthetic.

And what has been the outcome of these years of
experience, of revelations, upon myself ?   I can easily
assure you that I am able to go into a sick-room with
a feeling of assurance, *doubly sure that the battle
against disease has to be wholly fought by Nature*,
with my duties wholly relegated to the keeping of a

clear roomy field for action, attended with such services as she can easily ask for, and are readily comprehended; and as her interpreter to all friends interested, to all hearts involved.

I have little need to become confused over what drug or what combination of drugs are or are not indicated, and no confusion whatever as to how the brain, the power-house is to be nourished during the heat, the smoke, the confusion of conflict.

For me, Nature cures all disease even as she does the wound and the broken bone. Hence I go into sickrooms with that confidence, that hope, that cheer and happiness, that love for the work, incomparably greater than in the old days of mists, of clouds, of doubts; that were always attended with a feeling that no matter what my pains, there were never other results than those possible even with the most stupid, reckless, spoiling hands.

And I can go into sick-rooms with a feeling that my services to the sick are attended with a relief to severity of the symptoms, and a shortening of the duration of disease, that cannot be charged to a change of type or a run of luck; for seventeen years is a long time to be so favored; years which have behind them eleven years of previous experience to draw from, by way of comparison.

Readers, I now leave these questions with you, with a hope that you will at least think and reflect as you have never done before. My evidence is only the evidence of one man's experience, but so far as it goes it is scientific as to the thoroughness of the methods involved. You have not been called upon to believe anything dogmatically asserted. I ask you to believe nothing, nothing for me; believe only what seems to be proven, and consider what is suggested.

If the suggestion that the several tissues of the body constitute the only bill of fare for the brain when the conflict with disease is going on, becomes accepted as a physiological fact, it will never suffer any evolution through the ages yet to come.

In my next lecture we shall begin a study of the science of nutrition as applied to the culture and maintenance of health.

I shall try to show you in these lectures how we are to so aid Nature that you can grow into health even as a growing, a living crop can be cultured into a more prolific fruition, by a higher method in culture. And as your eyes become opened with the keenest of interest in the possibilities of the culture of health, the more closely you will see that attacks of diseases are the improvidence of avoidable cause and not the providence of God ; that there is a science involved in the culture of disease, no less than in the culture of cure.

PART II.

EVOLUTION OF THE BREAKFAST TABLE.

Woe unto thee, O land, when thy king is a child, and thy princes eat in **THE MORNING**:

Blessed art thou, O land, when thy king is the son of nobles, and thy princes eat **IN DUE SEASON, FOR STRENGTH** and not for drunkenness!— Ecclesiastes 10 : 16 & 17.

LECTURE XI.

"BEHOLD, HOW GREAT A MATTER A LITTLE FIRE KINDLETH!"—ORIGIN OF THE EVOLUTION OF THE BREAKFAST TABLE.

My Friends the Readers :—

Some years ago as I was walking my way along one of our streets I chanced to meet my friend of many years, and former teacher, Prof. Samuel P. Bates, A.M., LL. D., war historian of Pennsylvania.

He was carrying his head somewhat adown, as if heavy with thought. On meeting him he at once gave me the thought of his mind. He had just returned from an extensive tour through the old world, with eyes open ; and among the things of interest observed, was the eating habits encountered. He had on his mind a lecture on the subject of diet. All that he told me then was that upon the excursion steamer and at all the hotels the only breakfast was a cup of coffee and a roll, very small, I thought, for an American breakfast.

Now I think I had simply heard of this before, but here was a direct reception of the fact. Poets tell us about the " rapt porch of the charmed ear," of hearing with the " spirit's finer sense."

This kernel of suggestion dropped into a porch that was not rapt, and remained there I know not how long, without involving any of the spirit's finer sense. I simply did not forget that it was there, even though the lecture was never delivered. It proved to be a

143

very live kernel, but for some years the germinating conditions were absent.

I told you that I, myself, was a victim of that hydra-headed monster, dyspepsia then so called ; and I had a case that the old trite suggestion, "Physician heal thyself," was annoying to encounter. Briefy stated, I was the victim of slow digestion, never attended with pain but always with discomfort, which invariably disappeared under the excitement of business. I never was without appetite, in fact it was always too strong to be manageable. I was never disabled by it an hour in my life, but there was always an abiding sense of discomfort when not under excitement. Heredity supplied me with the constitutional conditions and I, duly and habitually, all the rest.

By virtue of a slight physique and weak digestive and assimilative power, and a highly-wrought nervous organization that never would permit repose in an arm-chair, unless asleep, vital power, by undue and always hurried eating, was kept continually overtaxed, and I had the toughness of muscle fiber, and the absence of organic disease to endure it, with the habitual consequence realized.

I habitually ate a hearty breakfast every morning with or without previous exercise. No meal ever fully satisfied hunger ; could take a lunch between meals with relish only that I dared not ; and all through the days, the months, and the years I was more or less starved because of habitually overtaxed machinery.

There was nothing to be found in the books for me but tonics to raise digestive power, and pepsins to aid as solvents. My abhorrence of the bitter, and respect for the integrity of the mucous membrane of my stomach, coupled with an utter inability to see how they could possibly be curative, and withal the manifest failure of such means in the hands of others, prevented me from dangerous experiments in this line, and from suffering, if not from "many physicians," certainly from many medicines.

In due time the effects began to appear upon the lines of my countenance, upon my expression and color, and, through a noticeable degree of mental apathy, as I appeared upon the streets. Friends thought I was becoming indifferent, would wonder at times why I should pass without even noticing them—and at other times with my attention engaged, I seemed to have lost in social liveliness. Very often I would be called upon to reply to an observation that I was not looking well, in fact that there was even a haggard look—and what was the matter?

The matter was to me that I felt every way as badly as I looked, and that to be a walking index of my mental and physical condition was only less annoying than the fact of its existence. In my case there was no settled gloom, no tendency to melancholy, nor any feeling of discouragement as to my human affairs; there was only the sense of discomfort, a feeling of inability, exhaustive in its effects, to do anywhere near the work my tastes, my ambition always kept before me.

When good John Bunyan first began to preach he was sorely tried by the temptations of the devil, who would insist at times on enforcing his company even to the pulpit steps, there to part with him until after the sermon.

In my case I could always throw off my evil one in the room of the sick, so as to have even the most perfect use of all my faculties; or whenever under any pleasing excitement. My sense of physical comfort would be perfect as the use of my faculties would be

10

strong. But between the scenes there would always come that sense of exhaustion as if I had been unduly stimulated by some abnormal excitant. This condition of excitement of mental powers has been aptly termed, when excessive, as a "dry drunk." Well, my sense of exhaustion was always such as would naturally follow such inebriation.

Between the scenes my "storage battery" had always need to be recharged. This starved condition of the body, this exhaustive tax on vital power, through an overtaxed stomach, always tends to mental apathy as well as to irritability of temper, and that my home did not become a "bear-garden" was due to the fact that my inner life was continuously under the soothing, cheering, uplifting influence of one who shared with sympathetic interest in all the weal or woe that life bestowed upon me.

There are the "slings of outrageous fortune," and there are the slings of outrageous tempers, also the slings of outraged tempers. Anger is emotional insanity; it deprives the mind of its judgment and the morals of their sense of justice. In its frantic, its furious efforts for relief, consequences are lost sight of; its revenges are always excessive. A fling is given to an excitable temper; a blow follows: Dr. Parkman falls and Prof. Webster hangs.

In my own case, as I began to see more and more the effect of cheer upon digestive power, so did I begin to study the virtue of repression for the good of those around me.

Said an old friend to me one day, "Doctor, I believe it is every man's duty to make his home as happy as he possibly can; I make mine happy by keeping out of it all I can."

There was a good deal of philosophy behind that remark that he wot not of. Even one cheery mind in a home is as the draft to the flame in its uplifting effect upon the physical, the mental and the moral lives upon which it shines.

Behind my life was the irritation of imperfect digestion, and hence I had need always to keep abnormal emotion in suppression when it seemed most to need vent. It is always a relief to the individual to express the emotion within, whether by a frown, a red-hot word or by a brickbat.

Mark Twain tells us that, while learning to run boats on the Mississippi, his irascible captain was about to engulf him in execration because of some oversight, when the mind was diverted by a slight collision, whence a furious bombardment between captains across gunwales ensued. When the powder was all burned, and when the captain turned to attend to his case, he was empty ; not the finest skimmer run through him could have gathered an oath ; and with the utmost mildness the needed suggestion was made. There is philosophy even in this exaggeration. We selfishly give vent to our irritations that we may find relief, but which always is at the expense of the cheer of those around us. By studied, persistent effort I found that I could often find relief by the slower process of diffusion or absorption from within, but which involved the virtue of silence, which was never like a "poultice" for the heat, the inflammation within.

In the course of my human events I arose one morning in unusual exhaustion from digestive taxing of an evening meal that was not needed, and seated myself at a breakfast-table where the supply was adequate for a noon-meal at a farm-house. For once I was not hungry. Why should I have been ? I had only had

the exercise called out in dressing since I retired the
night before ; but it had been thirteen hours since I had
eaten my last meal, and not to eat until the next meal-
time would involve an additional fast of five and a
half, in all eighteen hours and a half.   Should I take
the chances of going without a breakfast and be likely
to faint dead away while trying to count some irregular pulse ?   Preposterous !

There came a sensation in an' ear, a pouch, and the
suggestive kernel dropped in by my friend the Professor, dropped out—I resolved to chance a foreign
breakfast, but I was not hungry, and should I take the
roll ?   " Oh, you must," says habit ; says custom, " It
is the most important meal of the day after so many
hours of fasting, so you must always eat whether
hungry or not."

I took my chances on a breakfast of coffee only, and
so far from fainting during my forenoon ways, *I had
a forenoon of such lofty mental cheer, such energy
of soul and body, such a sense of physical ease as I
had not known since a young man in my later teens!*
When the dinner-hour came there was an added relish
that was a new experience, and I left the table with a
stomach so supplied that there was no need of apprehension as to an attack of faintness during the
afternoon.   But though that dinner was easier handled
than any I could remember for years, the afternoon
was not so marked with easy mental play.

The next morning's breakfast table was approached,
and, readers, do you suppose I debated the matter even
one second as to whether I should try the experiment
again?   I did not even consider it ; I simply imbibed
my coffee, and there followed another forenoon of comfort, of mental and physical energy followed by the
better relished dinner and duller mental conditions for
a time thereafter.

My evening meals, while always light, were always heavier than I had any right to try to dispose of, and these were not so well handled as were the dinners, because always too near them. In the course of a few weeks there was such a quickening of my life in every line, that friends began to notice it. And it was very cheering to be met on the streets with the remark that I was looking better; and all the more as I had reached a time when mental and physical wreckage seemed not so very far off.

Now, up to the morning of the first coffee breakfast, which may have been twelve years ago, I have no record of the exact time, there had never been observed anything in the play of my eyes, in the expression of my countenance, in the co-ordination of my ideas, in the conduct of my professional services, or in the management of my business affairs to suggest lunacy or crankism. I had been able to get my sick through their illnesses without the reputation of starving any of them.

Now, this going without a breakfast had a suspicion, if not of lunacy, certainly of crankism—for even I could not suggest one plausible reason why I had been so much benefited by it. The blind man got his sight, that he knew, he was certain of it; and not for him was the question of its philosophy, the how; he was only too glad to accept the fact of the new life and sunshine. And so was I—I accepted the fact and let the philosophy develop in its own good time.

I was now to enter upon a new career of professional services that was to involve a quickening of other lives even as my own had been quickened; to involve a study of the science of nutrition in the cure and prevention of disease, such as was entirely new to me, and yet which was to involve a method of living so in reverse of the

established modes as to only invite ridicule, denouncement, epithet and every possible mode of the expression
of disapprobation. But what was good for me was
good for others, perhaps—the remedy at least was safe.
I could no more tell why it was helpful than I could
explain so as to make it entirely clear to a lay mind,
just how a dose of an alchoholic tonic was to strengthen
the stomach for its habitual avoidable overtasks.

So I began to advise patients suffering from woes
innumerable arising from overtaxed stomachs, to
abandon all food in the morning except the cup that
"cheers but does not inebriate." This advice was
absolutely safe, because at most only a faint feeling
before the next meal could be involved, and if hunger
should become very persuasive before the midday meal
or luncheon, as a symptom of need, it would care for
itself. The supplies would not be far off.

Faintness is not always attended with danger; and
then was it not a matter of history, is it not, that in
cases of enforced fasting hunger has been attended
with a physical energy that would permit the procure-
ment of food even by stealing it? Is it not a matter of
history that, after a famishing fast, there has been the
physical energy still left, to slaughter a companion that
the survivor might live? Why not then absolutely
safe to enjoin a fast of only a few hours, even after a
night of rest and of fasting, with the bread of life
always available in case of pressing need, before the
regular eating hour?

I began to so advise, and they all who were advised
began to have their lives quickened even as mine had
been. And more, they began to advise others to go
and do likewise; and they who were most benefited
because the need had been greater were most enthusiastic over the way they were being saved.

In course of time a case came to me, driven because of the ceaseless importunities of a friend. It was a young lady from the highest social circles. During many years she had eaten all her meals as a sense of duty, and not from pleasure or hunger. She had appealed in vain to medical science ; she had tested the numerous highly advertised specifics, and the outcome was that her luckless stomach had reached a condition that prohibited all solid food. For more than a year her life had been sustained by liquid foods, such as broths, soups, meat, tea, etc. She came to me hopeless of relief and only to silence the importunity of her friend.

Without fully seeing the physiological reason behind the method, a strict fast was enjoined until there should be developed real hunger, and then to begin with the solid food that was most indicated. I succeeded in making such an impression that she left the office not a little enthused, and all the more as dosage was not involved.

This was my first marked case. This lady entered upon her fast, and, by a due prolongation, was able to handle a light solid meal with no painful efforts of digestion as formerly, and by studied care, she became in a few weeks able to eat two substantial meals daily without discomfort ; and in time was able to say, " I have no words to express what it is to me to eat what I will, with keen relish, and not have to suffer for it."

Her life became quickened in every way, her intellectual force became increased, and ultimately her learning, her culture and her power over the expression of written thought began to adorn the pages of our high-class monthlies. But I was not able to suggest a plausible theory why the inevitable breakfast should be abandoned—nor was she able to meet her friends with arguments, who were astonished, even confounded, over a change that presumed to come through a method that seemed to deserve not consideration, but ridicule.

This case caused a good deal of discussion in the city, but it gave me more of ridicule than of credit; that such an end should be achieved by such means—well, it staggered belief—and because it had not behind it the mysticism of dosage.

The Syrian leper was simply disgusted with his prescription; he wanted a miracle and not a succession of baths; and the Jordan was inconvenient, and no purer than Abana and Pharpar.

In the case of this lady there were no associated conditions of disease that I became aware of, and there was no actual disease of the stomach; it had been kept only functionally overpowered year after year. And medical science had only been directed to a result and not a cause, and hence it inevitably failed. There was only the one visit to my office, and afterwards the chance meetings upon the street. And as to the method, she informed me that she was able to say to all, "As long as I keep in line with the method, I am all right, and when I do not I am all wrong."

For a long time after I began to advise patients in this way, I saw nothing but the stomach, structurally or functionally disabled, to be relieved, and with only general improvement to be expected. But, as the scope of my vision began to enlarge and to clear, there began to be results noticeable that needed to be explained, that must be explained, and the very first that met me was a loss of weight.

LECTURE XII.

My Friends the Readers :—

My attention to a loss of weight, while a cure seemed to be going on, was incited by a case that caused a great deal of discussion in high circles in the city. It occurred in an infant of four months that seemed to be fat, of full weight and well nourished, which had had a bowel trouble from birth ; and, though not supposed to be sick, yet had been under light dosage every day of its life. It was the first-born of a young mother who was capable of convictions : and who had all the needed resolution and perseverance to carry them out.

By some chance a friend sent her a little book from Boston written by a Doctor Page, which advised only three feedings per day for infants—three nursings in 24 hours, only !! And the next day after reading this advice she began with three nursings, six hours apart, rigidly carried out.

In less than two days the bowel trouble ceased, and in every other way the improvement was marked ; but it was not very long before there was such a decline in weight as to cause some uneasiness to the parents, and a hurricane of disapprobation among the army of relatives, because of the starving method ; and I was called because of known views in some harmony with the condemned book. After a careful history of the case before and after the reform-diet method was

153

adopted, I could but believe that the waste that was manifestly going on was not revealing any danger symptoms; and so I advised the method to be continued, and hence put myself within the storm circle as an aider and abettor of criminal stupidity. I watched over this case for some weeks, and saw that every attempt to interfere with the regularity of the six-hour meals, was followed by an immediate digestive disturbance.

An additional adviser was called only to be confounded over the result, but was not able to advise a change of the plan. We all had to accept the fact that the loss of weight was in no way a process of disease, for the child grew into sturdier health that was steadily maintained, and physiological law triumphed over unreasoning prejudice. But why or how the loss? That was beyond my reasoning out, for a time.

Next came the case of a young lady in close social relations with my family, an only daughter. She was advised to abandon food in the morning, not because of disease, or apparent weakness of the stomach, for of this she never complained; but because of a feeling of languor, of inability to attend to her affairs with any energy of body or mind.

She was finely formed and with not an ounce of "flesh" to spare for harmony of proportion. In the course of a very few weeks there was some alarm all through the house because of a loss in bulk, and I could not account for it; the clothes were becoming too loose for due fitness. But here was another fact, the general health had become so improved that she could go about with elastic steps, and with an energy and cheer that was a revelation to all; in fact she could run upstairs without getting dizzy and out of breath, whereas she could only walk slowly on this account before. How was this to be explained?

The clue came for a solution by a chance recalling of the bloated, dropsical feet and limbs of old people, and those of broken and weak constitutions, who have this difficulty to contend with after illnesses that have confined them to the bed for a long time. Here we have thin blood, thin, lax blood-vessels, the elastic or contractile fiber in the vein walls so stretched by loss of tone, as to permit the veins to become dilated into elongated sacks ; and these sacks have to hold their contents against the force of gravity.

In this sense there might be put this appendix on many a small headstone, in the cities of the dead : "*Died from unrecognized cruelty.*" And it would be true in a frightful percentage of instances.

In the case of the young lady, the loss was accounted for in the same way, a long time after apprehension had ceased.

Years after this I had a most striking illustration of the soundness of this theory of loss of weight. I was called to take charge of a case of general and abdominal dropsy in a farmer, past his 60th year.

Very learned counsel was duly called, exceedingly skilled in the diagnosis of disease, and who carried in his mind a wealth of medical lore. He exhausted his powers in a search for the cause without getting a ray of light. It was his opinion that there was not less than a gallon and a half of water in the abdominal cavity. His suggestions as to treatment did not prevail with me, because too destructive of digestive power, and then it was only aimed at a condition, at a symptom.

For weeks after, though the patient was able to be about the house, and the sleep was not interfered with by any difficulty of breathing, yet the case seemed be-

yond all tinge of hope. But this man's stomach was
guarded with a care, with a vigilance equal to its
importance as the last, the only foundation upon which
to base any possibility of relief. Readers, this man
remained in his home an entire winter, very bravely
facing the possibility of the grim messenger—but the
following winter he spent very many stormy, cold
days in the woods at one end of a cross-cut saw!
Richer blood through better *acquired digestive conditions*, thicker-toned blood-vessels, and absorption of all
the *dislocated* water—that was all ; that is the whole
story. There have been three years since of health
such as he had not known for many years.

I came in contact with this patient at my entrance
upon general practice in a way that made a striking
impression upon my mind as to the need of relieving
the mind of apprehension, when we are trying to cope
with disease, and also of the necessity of so construct-
ing the language of the thought of our minds in a
way that makes it easy to vitalize. I had known this
patient all my life. In his later teens he had been told
by a pretender that his heart was diseased, and, possessed with strong imagination, he had gone on through
the years with the dread feeling that a sudden death
was his possibility at any moment.

But full of ambition, with an elastic nature and
with muscular agility that would have made him an
expert in a circus-ring, he failed to suddenly die. On
my return from the army, I found that for some
months he had been in a low mental condition through
apprehension that death was not so very far off, and he
lost no time in getting to me with his sad story. I
found no trouble with the heart, but I could not relieve
the apprehension, for the heart trouble was a *fact* in
his mind. I failed to have just the logic I needed,

until there came a time when a fluttering and an actual soreness over the region of the heart roused him as never before. To him, this was evidence that was beyond any question—a sore heart was not to be relieved by logic, so it seemed to him. He mounted his horse and rode two miles to have his worst fears confirmed, to receive from me a sentence of death !

He told his story, and I addressed him about as follows : " You really believe, then, that this soreness indicates a sore, and therefore a diseased heart ? "

" Yes."

" Well now, see here, just think a minute ; your heart is all muscle, and a very strong muscle too ; do you think it can be possible that a sore heart could beat, contract, relax, without agony any more than you could bend your wrist joint if there were a boil located upon it without torture ? Again, don't you see that your heart is not lying in pressure against the front chest wall, and that hence you could not feel a soreness by pressure, *through space ?* "

Readers, that man went home with that heart which was in his mind, so light, that from that day he never again mentioned his heart to me. The heart was cured through logic. Had I been unable to relieve his mind, his apprehension, by reason of its depressing power upon the forces of life, would have paved the way for an earlier attack of that disease that seemed so ominous of danger.

There came another illustration. A man supposed himself the victim of a heart-disease, because of attacks of fluttering. His wife had got a hint of it, and hence she was full of apprehension. Now he went on enduring this for some years with slow declining general health and an aggravation of symptoms, but kept his condition concealed from the wife, as far as he was able, to keep her out of trouble ; and she the while not daring to hint to him as to his condition through fear of getting a fresh addition to her own apprehension. He dared not consult a physician through apprehension of having his fears reinforced. At last the strain became too tense to be longer borne.

He came to my office in sheer desperation, determined to face the worst ; he rushed me into the consulting-room with a drawn, grim expression, and even without a hint of what was wanted of me, bared his breast, and with forced utterance ordered me to *examine his heart.*

After a lecture on the "science of nutrition" he returned to his home clothed in his right mind, to enter upon a higher physical life, and never more to suffer with apprehension from disease of the heart.

The woods are a luxury of wandering during the sunny hours of the day, but at night when all is the darkness and silence of the dead, how drear, how possible to throng them with imaginary dangers ! Who is so brave that he can endure a " haunted house " for a night, without a shade of apprehension ? Apprehension of the possibilities of disease is the "haunted house" of the nights of our lives.

In entering upon this new career of medical method, the treatment of so-called "chronic disease" that had been practically ignored for so many years, my office was to be converted into a lecture-room, where ideas, that were being vigorously antagonized by denouncement, had need to be put in the clearest, the simplest, the most condensed form of expression. And this need led me incidentally to a new interest in the study of mental science. It gave me a large opportunity to study the receptivity of minds in a comparative sense ; in other words, the relative capacity of minds to receive and vitalize ideas that are capable and worthy of vitalization.

And I am able to say to you that, after many years of experience of daily opportunities, with the mind intensely concentrated on the receptive symptoms, I am just as uncertain when, for the first time, I address a Doctor of Divinity, a Master in Arts, as to the receptive condition, as I ever am when addressing a laborer or mechanic.

A receptive mind which can absorb truth to enlarge the mental power-house, is an endowment of nature. The schools, the colleges can no more give it, than they can give poetic genius and inspiration.

As the years have gone on I have learned to estimate minds as moral and mental power-houses, and not as ware-rooms or academies of fine arts. It is the mental power-house in the machine shop and in the " sequestered ways of life" before whom I bow with lowest reverence, because it is to such we owe so much for the advancement in all the material interests of life. We have our lesser Whitneys and Edisons in the commonest walks of life, without being aware of it themselves.

But why the attack upon the breakfast-table of my patients? Well, one very decided answer all patients could make was, that they always had more physical and mental energy without, and consequently a larger degree of comfortable feeling, than with the breakfast ; hence there was no disposition to go back to a habit that would simply invite trouble. In the course of time, as there was a constant addition to the ranks of believers, so the pressure came upon me for a solution of the question which would satisfy the science behind it, if any were to be found, and all the more as

my professional brethren were beginning to comment.
One went so far as to privately admit that on this subject of going without a breakfast I must be "absolutely crazy." His intellect was as keen as a "Dam-
ascus blade." Physicians are the conservators of the
public health, even against the dangers of a medical
heresy with a lunatic behind it. My comfort was, during those earlier times, that all opposition, no matter
how couched in ridicule or sarcasm; no matter
whether from a professional or a lay mind, was aimed,
not at me, but at a larger compliance with the immu-
table laws of life itself, and hence I could bide my
time and did not allow my mind to become soured by
opposition, no matter how intense or in what form it
came. The physician never saw any change in me,
nor I any expression of his countenance or glance of
his eyes, when we met, that indicated any apprehension on his part, that a sudden outburst of incoordi-
nated mental conceptions were possible, or that he was in
any danger of violence thereby.

We will now look into this matter of going without
a breakfast or rather of postponing it to an unusual
time, to learn, if possible, whether it is based upon
science, or a whim. The first suggestive clue to the
solution of the problem came with the thought that
sleep, as compared with violent exertion, is not a
hunger-causing process.

The evening meals are taken and rarely fail to be
more than ample for all the destructive need from cell
action that ever takes place before entrance upon
sleep. In perfect sleep the self-generative cell changes
are at the slowest as they are the most rapid in the
most violent exercise. These changes go on with such
precise movement that the breathings are regulated
with a like precision. They are just so far apart, and
are alike in slowness, length, depth and force, all

adjusted to the need of oxygen. Now, sleep and rest, after severe exertion, ought to be refreshing.

> " Blest be the man, said he of yore,
>   Who erst the famous Quixote's target bore.
>   Blest be the man who first taught sleep,
>   Throughout our wearied frames to creep."

There is the night then of perfect rest to muscle and mind. Does this refresh? You should be refreshed by this experience during which you have been oblivious of all care or woe of mind; or are you so exhausted by it, that you cannot go about your affairs without a filling of the stomach? What is sleep for if not to refresh? When should we be so able to go about our affairs as when this refreshing has been most perfect in all its processes?

Readers, listeners, *there is no natural hunger in the morning after a night of restful sleep,* because there has been no such degree of cell destruction as to create a demand for food at the ordinary hour of the American breakfast. *Sleep is not a hunger-causing process.* To reinforce this statement and the reasons behind it, is the experience of thousands who have abandoned the morning meal, and in a short time lost all hint of a need of it. This could not have been had there been a need, for Nature is imperious, exacting; and it is not in the line of possibility that she will permit any getting used to less food than she requires to preserve her physiological balance. She easily permits you to skip that meal you do not need so soon after the refreshing sleep and *which you always eat from habit;* but later she will call you to account if you give less than her demands.

I will now leave this most interesting point in our discussion until I can go back and bring my thread of history up to where I can make an attack upon the

11

breakfast all along the line. Thus far I have been content to advise the "sick and afflicted" only.

I must tell you of an evolution that took place at my own breakfast-table. As there were no apparent reasons why any of my family should adopt my plan, not a suggestion was offered to any of them that it should be done.

During the winter following its adoption, I sat at the morning table and sipped my coffee while my three sturdy boys enjoyed their steaming cakes fresh from a smoking griddle, with such unction of relish as to always make me feel like sighing with regret that I could not be a boy again with a boy's powers of digestion. These boys would begin to break their fast on this kind of "bread of life," and end it with the same, for no other food was wanted, nor was there room for other kinds. The last mouthful was the cork to fullness complete.

Now it had not escaped my attention that this was the only relished meal of the day, the others being encountered with a dainty, eccentric appetite, and that the one morning meal of the week when they were not hungry was on Mondays, when there were no hot cakes to entice stomach-packing. On these mornings the three of them would not take as much of the less enticing food as one would dispose of on all other mornings. This did not prevent their appearance at the dinner table with no lack of mental or physical energy for the task of eating such a dinner as was eaten on no other day of the week.

The following winter, at my suggestion, this enticing food was made the second course of the noon meal in order that there should be due variety in the bill of fare, as well as to avoid the taxing of the stomach with so much food of difficult digestion. On the start there would be the advantage of not over-taxing in the morning, as the morning meal was so plain as to prohibit possibilities of gluttony. This plan worked like a charm. The breakfast at once began to decline in interest. Without a hint of suggestion the wife got down to the foreign breakfast, and the boys began to be content with " bread alone." Occasionally the wife would be content with coffee alone, and then she would get a headache, and add to the next morning meal. Occasionally the oldest son would skip his morning meal entirely, but he was always on time for his dinner to restore the lost balance. And the second boy gradually cut in his breakfast.

As this evolution went on so did the general cheer of the entire family increase. There was a luxury of relish in those dinners, not only on account of the absence of eccentricity of tastes, but in the higher social cheer that most abundantly rewarded for a fast that was without ever any taxing of endurance or even a hint of starvation.

And it was also very noticeable that apples, pears and peaches, that are " so healthful," that they can always be eaten with impunity, had suffered a material loss in their tempting power. They did not care for them in the forenoon, because the light breakfast so satisfied every physical want that there was little temptation to partake : and the noon meal was so enjoyed and so thoroughly digested as to prevent a morbid craving for the acid fruits which are always considered so healthful, that they can be made an additional load to any stomach, no matter how overloaded already, and with no evil results.

With higher digestive power, and the habitually relished dinner there resulted such habitual freedom from morbid cravings that all fruits, no matter how healthful, lost their power to add a small meal between the two regular meals, and hence resulted robust, sustained health.

This evolution went on until the breakfast abolished itself, and without any hint that it should be done from higher authority. Except in my own case the frequent taxings in nights of professional care made the morning cup as a gift of the gods. The sons were never permitted to acquire the need of tea or coffee.

There was always the same self-abolished breakfast in the kitchen and always by evolution. No kitchen-maid ever served who did not of her own free will get down to an ability to do the hardest forenoon's work of each week with more ease, power and comfort, with a stomach void of food, than ever was done with a breakfast in it.

As for the sons they were always able to cope with their fellows in all taxing recreations, no matter how severe, with never a hint of faintness ; and were always able to approach the dinner table with the deliberation of old men. There were never any symptoms of nervous, impatient, exhaustive hunger, and they never failed to sit by until such meals were eaten as an empty, rested stomach only ever invites. This regular experience opened up the question of difference between normal and abnormal hunger ; and it opened up a most interesting and important fact, as satisfactory as important. A fact that when those boys were able to put the relished first meal of the day into the rested, empty stomach, I need have no apprehension of attacks of disease for an indefinite time.

Now, with my attention attracted to this evolution, with the highest possible human interest in its possible danger, and with such marked results for good, could I do other than advise others to "go to" and do likewise ?

LECTURE XIII.

My Friends the Readers :—

Since we met last you have been doing some thinking. You took your evening meal of yesterday with
more care. For once in your lives you entered upon
sleep with an idea that it might be well to let the
stomach as well as other muscles have a rest during
the night. Your sleep was not so profound as you
expected because, there being no digestive work, the
brain was not affected by the torpid influence that
attends the digestion of a large meal. But the sleep
you all got was so restful that you awoke refreshed,
and you decided that after so many hours of rest and
sleep you ought to be refreshed and not hungry ; and
though at the usual time you had an attack of habit
want, that seemed to indicate a hot breakfast, you resolutely met the attack without yielding, and so, for
the first time, have appeared before me with empty
stomachs, and will therefore be able to appropriate
whatever I may offer that can be vitalized, with more
power of reception and retention than you have ever
realized at former lectures.

You entered upon this morning's fast with a decided
advantage over me when I began mine, in that you are
aware that it is absolutely safe, so far as the integrity
of the brain is concerned : that you could prolong it

until the skeleton condition would be reached without danger to life.

You have already gone past the morning attack of habit hunger, and this naturally confronts us with the question of hunger, of appetite itself. I promised to enlighten you on this subject in the first lecture.

In medical practice the question of creating appetite has always been a perplexing one. There are various remedies that are used for this purpose. A few years ago a very learned physician informed me that powdered Peruvian bark was one of the best of medicines to create hunger and tone up the system ; and, as evidence of his faith, he was then taking it for the reasons assigned. I at once began to test its powers as an appetizer, and to use it without violence to the sense of taste, or injury to the stomach. I prescribed the impalpable powder with the doses incased in capsules or wafers ; and for a long time I thought I aided Nature in this way, but the evidence was never inspiring, and then the looks and size of the dose savored too much of the dark ages of medicine.

When I tell you that in a new work on the practice of medicine that I procured because of the eminence of its author, and because a desire for the latest news on the subject of diseases and their remedies, you will be surprised to know that in cases of chronic lung diseases, where the appetite has failed, a sea-voyage is advised, and if this fails, then enforced feeding becomes a necessity !

Hunting for an appetite on the seas, at the ocean-side and among mountains, is a fruitless occupation with very many invalids ; and thus far medical science has failed to suggest a less expensive and more effectual means as a last resort.

Barrels of cod-liver oil are daily taken in this country because of results that are clearly traceable to habitual eating without hunger. In my own city physicians have their seasons of energy in pushing this taste-defying lubricator, and the chemistry of pharmacy has been at work for years trying to so disguise, as to permit an easier slipping past this life-guard of the stomach, but all to no purpose. The artificial appetizers are failures; and so the search must go on and will go on, so long as the natural means is unknown, and the need to feed in order to "keep up the strength," no matter how absent the natural want for food, is deemed a necessity.

But how sudden the revelation to me! Go without your breakfast and you will be hungry for your dinner! And so hungry that you will forget to take your cod-liver dose! And the dinner is so well relished, and you feel so much better after it that you conclude to omit the dosing altogether! *How simple!* Only a fast, no matter if it costs a whole day, a whole week or a whole month, and with absolute safety; why, do you not recall how energetically the digestive organs will work over the keenly relished food after the long fasts due to fevers? How much more then may be expected from fasts that are to be no tax on vital power! Safe? Yes, beyond any question. As soon as the stomach and appendages have disposed of the decomposing, unbidden meals that are still a tax on vital power, there will be a positive increase of mental and physical power, so that when Nature's own signal for food is given, there is none of the exhausted feeling that is more or less realized before the needless morning meal.

And the appetite will always come where death is not inevitable, no less in the ordinary conditions of low health than in cases of acute sickness—*and it is the swiftest, the most effectual, the most unfailing of all devices ever conceived for inviting natural hunger.*

We will now go back and see what is going on in the stomach while the fast is on. When you arose this morning duly refreshed, your system was at a physiological balance, the eating of yesterday having been expended upon the losses of yesterday, so you opened up a new set of books with Nature. You are going to get into her debt by a running account which will mature in your cases about noon or later, when you will get down to your table with a much pleasanter settlement than may be realized in a merchant's counting-room.

Now these stomachs were absolutely empty when you arose, and every moment of the fast will be a moment of developing gland power, whereby the solvent juices are to be thrown out in floods upon the relished food. There takes place in these glands, in their functional development, just what takes place in the cow's udder after the morning pail of milk.

There will be also a corresponding development in the muscle energy of the stomach whereby the food is to be more swiftly and powerfully revolved, so as to press out and wipe off these powerful juices; in short every function of the stomach will gain in power by the rest.

The glands of the mouth also gain in functional power ; the sense of taste also develops whereby eating becomes a luxury of life. That this is strictly a physiological evolution, no writer will care or dare to go into print to confute. Every meal taken after a fast that has caused keen hunger is a demonstration beyond any question. The stomach can be fairly felt at work. Now comes the question, *what* shall we eat ?

Let us all go where we can get a sort of a "kindergarten object lesson " on the physiological science of dietetics. We will go to the dining-room of the axmen's camp, at the noon meal of a January day, and while the sturdy men sit down around their loaded table, we will stand around it to form the outer circle. Is Nature going to permit any time to be spent in investigation, before the relishing work begins as to the amount of starch in this food and the albumen in that, or as to the relative amounts in each ; shall we hear about the " hydro-carbons " and what had better be eaten as to the amounts of the salts of potassium, of lime and soda they contain ? Nature will brook none of that, for she is in a hurry to balance large accounts. No food on that table but the most " nitrogenous " kind, and when the last column is run up, the expression of countenances over accounts settled, is of the earth-heavenly in its supremest satisfaction. There is no question there of what had better be eaten, that question Nature settled herself before the first course was attacked. There are no enticing persuaders, for none are needed. Nature made out the bill of fare for each according to the exact need, and without the least aid from the science of dietetics.

And those heaviest of all days' works are done with no craving for the acid fruits between the meals that are so " healthful " ; and all with no care or relish for other than the plainest, the most substantial food at meals.

Is anybody going into print to deny that Nature is unable to make out her bill of fare, that she does not, cannot know what food is wanted to balance an unsettled account ? *Keen hunger, hunger only makes known the individual need.*

The sense of taste then, you see, as you have not quite realized before, exists for a two-fold purpose. (1.) To

indicate the precise food needed to restore the wastes
of muscle energy, and (2.) that there shall be no
mistakes made, the needed food is to be the most keenly
relished. Now with this to guide you hereafter you
will not need to study the science of food analysis, if
you so allow your appetite to develop that Nature can
order the bill of fare *out loud* with the clearest enun-
ciation.

It now begins to break in upon you that I am a very
liberal feeder, and hence do not deserve my reputation,
founded, of course, and, after all, naturally founded, on
misapprehension, that my patients are kept under
more or less continual anguish from want of the
"bread of life."

I never presume to make out bills of fare for my
patients. It is my business to so educate, that Nature
may speak with power. You now fully understand
my views on this question.

A second question and of almost equal importance,
how shall we eat? Well, if we have lost no time in
the morning over a breakfast, we should be able to
afford by that much at least more time over a dinner
that will more than balance, in its pleasure even, what
was missed at that hour. Now most breakfasts are
taken, not so much from any real want or pleasure, as
to provide against a want before the ordinary time of
the second meal—and so the hapless stomach is made
the vehicle, and the very worst possible, for *carrying
purposes* merely. No human stomach was ever made
for a lunch-pail to carry food in, before needed.

I have already, in a former lecture, suggested that
digestion begins the instant the first mouthful comes
in contact with the gastric juice ; chemical changes,
you know, are often instantaneous. Now it takes
time to satisfy hunger, and as the supply of gastric

juice is limited, as there is a limit of the milk that is to be drawn into the pail, can you not at once see the necessity of eating so slowly that hunger will be sated before you have exhausted the juice supply ?

Listen to this statement : *Every disease that afflicts mankind is a constitutional possibility developed into disease by more or less habitual eating in excess of the supply of gastric juice !!*

Eating slowly guards against this danger by toning the appetite down before the danger point is reached. Slow eating renders the food more soluble, hence more favorable to the chemical forces involved in digestion. Slow eating favors the digestion of the starch in the foods through the action of the saliva upon it. You have been told that digestion is a tax on vital power, that, during its active stage, muscle and mental energy is correspondingly deficient, hence the sluggish feeling you will always experience after your dinners, which will be unduly heavy for a time.

I have also suggested that cheer of mind is as the draft to the flame in its stimulating power upon this function. This fact cannot be over-estimated in its importance as an habitual condition to be provided for after each meal, so far as there is power to summon it.

Have you any more important business, can you have, than to aid the stomach in every possible way while it is engaged in balancing an account that is to prevent disease from entering a mortgage upon your life ?

Some of you have children; you will now more clearly see that the meal hour is not the time for lectures on moral reform, because you have an audience within your grasp ; you will postpone the digestive, depressing chills until cheer and the highest social pleasures have sent new life through all the lives about you ; and, after digestive balances have been duly struck, then the moral shortage can be more serenely considered. You are also to keep alive the importance of not approaching your first meal *unduly fatigued* by which your eating will, by so much, fail to satisfy the need.

Now a supreme advantage of this method of creating an appetite and of so ordering the times of eating, is that you will actually eat more food on the average than you ever did before, digest a great deal more of what you do eat. How is this, you ask ? You will do this by keeping your blood rich ; with disease, even with microbe, defying power. Your health will reach such a high average, you will be habitually so much stronger that you will miss very many less meals, because you will miss the ailments, slight or great, that probibit them. They who can eat three meals daily and keep it up month in and month out, are a small per cent. of the people.

Now you are to abolish your breakfast and never presume to eat again without keen hunger ; this hunger you may have if you wait for it, even while sitting in an arm-chair or lying in bed, and it will be for food as nourishing as the axman requires. What shall be eaten at each meal will be a law for self to determine. No food is good or healthful and therefore typical, without a special demand for it. *Keen hunger, the most relishing of foods, thoroughly masticated, a recreative state of mind during digestion,* these are the easily-acquired conditions behind sustained health.

The second meal of the day should be so light that it can be duly digested before entrance upon sleep. But this raises an unsettled question—is there need of digestion during sleep ?

Let us consider ; there is very little destructive

demand for food going on during sleep, the vital machinery therefore requiring very little fuel ! Digestion therefore, and necessarily, will be a very slow process. The lying posture is not favorable to the peristaltic sweep of the food around the stomach.

The brain can suffer no loss by reason of an empty stomach, therefore if it is not subjected to the taxing power of digestion, the sleep ought to be more perfect. It has been found by those who have adopted the two-meal method that a second meal unduly large is always followed by diminished mental and physical power during the following forenoon.

It has been found in repeated instances that if the stomach is permitted to rest during the night, it has a marked increase of functional power during the following day. From my own experience, and what I have learned from others, in connection with the fact that the brain will be duly cared for, I have no doubt that the stomach needs to be kept empty during sleep. This question is one of vital importance, but fortunately is one that can be solved by experience, and permanently solved.

You are now ready to ask, how about this method of living as applied to those engaged in manual labor ? I did not for a long time attack the breakfast table of laborers. In this as in all other respects the progress of the diet mode in the enlargement of the general health of the people, was a matter of evolution in which each advancing step that was made had a revelation behind it. As soon as I began to get people to abandon or scale down their morning meals, I began to get results ; surprises to explain, and that served to push on to other steps as I was all the more watchful for results because of the intense opposition that met my progress in every step of the way. Very often it appeared that my business annihilation seemed likely to be involved.

It is said that a certain Athenian philosopher once offered a theory as to the origin of the world that did not at all suit the minds of his fellow Athenians, and straightway they confuted his doctrine by banishing him from the city. The doctrine that I began to teach was, in the public estimate, a great heresy, and it had to meet Athenian logic without its power of suppression.

I once saw a powerful man bracing himself against a door. There was abject terror in every line of expression, and his face was streaming with perspiration, and every muscle was drawn to a steel-like rigidity, for his life depended on his herculean efforts to keep the robbers out that existed in a brain crazed with strong drink. And the uninstructed people braced themselves against this new "gospel of peace and good-will among all men." Indeed there came a time when my services at a very sick bed were involved somewhat in a defensive attack against the superstitious fears from without, because of methods that were deemed homicidal. Those were the old dark days when sometimes the pressure was scarcely endurable and I would fall back on this philosophy. "My own life has been saved and a new lease granted of greatly enlarged mental and physical force. I was becoming a wreck, the wreckage had only left its scars and memories. And my family were being benefited ; those boys were growing into manhood with health habits established, that as a heritage would be beyond any estimate in a pecuniary sense."

What is the rich man's millions to him as he sits down to his table with a stomach that will not hold a teaspoonful of food without a protest of agony ?

And then the happier faces I would meet on the streets, or in the brighter homes, would be like the fresh brigade to the wavering line of battle.

LECTURE XIV.

EVOLUTION OF THE BREAKFAST-TABLE,
(Continued).

My Friends the Readers :—

You are interested to know how the method succeeded in its attack on the breakfast table of the manual laborer. I am able to say that I did not begin the
attack. There chanced to come into my office a farmer
from some miles beyond my circuit, with a variety of
ills to open up to me. His complaints of rheumatism,
of stiffened joints, of the persistent cough and of
his broken-up, broken-down condition generally were
couched in all the language necessary to portray them
in vivid outline. He was beyond his sixtieth year and
of a most resolute, ambitious, persevering nature.

I was in a rather unusually garrulous mood, and so
I attacked him with a lecture on health culture, and
let him depart without a bottle. He was simply to
take a breakfast of coffee and hold off until dinner for
his solid food. I omitted, however, unintentionally,
to suggest that as soon as he became able to go about
his affairs, by reason of better health, he could begin
with the breakfast again, but that it must be lighter
than before.

This call was during the autumn. I saw nothing of

him again until after the harvest of the following year, when he came in to tell me that he had taken no other breakfast than the coffee since he was in to see me ; that he had never worked harder during forenoons in his life, and that he felt ten years younger, but that there had been a small loss of weight (he was of a muscular, wiry build).

"How about your rheumatism and cough?" I asked.

"I have nothing more of either of them."

Four years later I saw this man again, and found him more of an enthusiast than ever. He had abandoned his morning cup, had his dinner an hour earlier, the second meal at the usual evening hour, and on Sundays *only one meal*, with Monday always the best day in the week to work. And more, that during the four years, he had not had even so much as a cold, whereas, formerly, this was always a cold season annoyance.

Thus the power of rich blood to restore the waste places, and to defend against disease !

Speaking of rich blood, recalls my promise to tell you of the defensive measures I instituted against the bacillus diphtheria in the stricken family.

With one dead and three more on their death-beds, and seven more as possible victims, with the storms of November to prohibit exercise in the fresh air, with no other homes to admit any of them by reason of the intensity of alarm, the situation was simply appalling.

I at once had them all quarantined in a small but well ventilated room upstairs, from which they were called down every morning between nine and ten for a breakfast that was as hearty and as keenly relished as if after a ramble through the woods. They were called down for a second hearty meal earlier than the or-

dinary evening meal. The eating was to be entirely
confined to these meals, no piecings.

With me the subject of intense anxiety at every
visit was, the relish and amount of food taken by those
children. Those lambs of the flock continued to come
down, except the infant, *one by one*, but in " Indian
file," in an advance upon a *well-filled table*, and this
day after day.

After the alarm subsided some of them were taken
into other homes and subsequently suffered light attacks of sore throat, but not until after they were relieved of my starving method and put back upon three
daily meals.

In line with the experience of the farmer is that of
the brother, a sturdy farmer beyond his fortieth year.
He had been afflicted with more or less stomach trouble
for years, but I was not, in the earlier times, able to advise him to go without the supposed needed morning
meal because of his heavy labor ; but from time to
time he was made aware of results achieved by my
method, and he began to study the matter for himself,
and as a result there was an evolution of slow progress
that consisted largely in giving up fruits between
meals and the giving up of articles that he found were
an injury. He finally got down to a morning cup,
which, after a time, was given up, and then there was
a marked general development in health growth. He
soon reached a stage when he could perform such forenoons of labor as were never exceeded in the lusty days
of a matured manhood. And during these years he
added fifteen pounds to his muscle weight, and escaped
even so much of an ill as a slight cold. It is his present opinion that he could easily go on with all his ordinary work on even one daily meal.

Again, one year ago, a frail man who had been sub

12

jected to attacks of the whisky habit frequently during more than twenty-four years, whom I could never
get into my power long enough between attacks to
help him to his feet, moved out of the way of temptation to the country. He there rigidly abandoned his
morning meal, and when the season opened for heavy
work, he was able to go through all of the many unusually torrid forenoons with no hint of any physical
need until the near approach of the first meal ; and
he assured me, in emphatic language, that he went
through those forenoons with far more ease, vigor
and cheer than he ever could in the afternoons, with
the stomach engaged.

Of course the reason of this was a profound mystery
to him. But you all see that this could easily be, with
nerve force in co-operation with his labor, and no loss
by the diverting tax from stomach work.

At this time there is a carpenter walking every
morning to his work, who, by reason of a very weak
stomach, used to give me a very large patronage.
Once he came to his bed, and for weeks death seemed
inevitable, because scarcely water could be borne—but
he escaped with his life. Later on he became the subject of diet evolution, and therefore worked out his
own salvation. His health had always been frail ;
but now, past his fortieth year, when he ought to be so
broken up as to need my services more or less habitually, he is giving me nothing whatever to do, and
he does not reveal in his looks, in his appearance, in his
gait, the slightest need of my services. He starts out
with a cup of chocolate in his stomach, and a lunch pail
in his hand to walk a mile to his labor, and all day
long he does the work of his hands with a power,
spirit and ease he never experienced before.

How could such revelations other than incite me

to more vigorous attacks upon the breakfast table, no matter how heavy the manual labor to be performed?

For some years my attention was only called to the general improvement realized, with local disease a " side issue," and it was only when patients themselves began to notice improvements in their local ailings that I began to consider the possibilities of the new treatment in this respect. There were catarrhs, nasal and bronchial, that had stubbornly resisted all the science of sprays and douches, that absurdly began to improve as soon as digestive power began to develop. *And how and why?* The solution was a matter of evolution; it came, if I have solved it, in a gradual way. I will illustrate by a case that revealed a decline in several local ailments, which will also furnish a striking illustration of the causative versus the symptomatic mode of treatment.

The patient was a lady not far from fifty years of age, of queenly form, in fact, every way so finely proportioned and of such full habit as to indicate good health, and she was so environed as to always be able to admirably reinforce any treatment instituted. She was the victim of mentally depressing, cheerless, general debility, associated with rheumatism that so affected her hands that they could hardly be closed in the morning; and the limbs so that it required more or less effort to get locomotive conditions established. There was also a weak heart with a degree of chest soreness in its vicinity which, combined, were a source of persistent annoyance. And there were also the uncertain and not infrequent attacks of neuralgia of the stomach. And, to sum up, there was also an utter absence of appetite except the spasmodic hunger, that comes to the front in the presence of the viands of a well-filled table. For more than a year she had

approached the table, except when debarred by the
neuralgic attacks, three times daily with indifference,
and left without refreshment.

Now these various symptoms had been treated as so
many different diseases, which, by some inscrutable law
of disease action, had become separate, distinctive en-
tities in the location of their choice, and each was to be
attacked and dislodged by specific remedies, and not
as the direct result of the action of one primary cause.
"Similia similibus curantur" had exhausted its
powers on these symptoms with not the shade of relief.

After a great deal of persuasion this lady permitted
an interview with the enthusiastic apostle of the new-
born heresy ; and she was surprised to learn that her
stomach, which she did not suppose was ever at fault,
except when under throes of neuralgia, was the source
of all her woes ; and she was more than surprised to
learn that a fast was to be enjoined until keen hunger
would come, if it required days to bring it about, and
surprised that such a method had never been tried
before.

She began the fast with some hesitation and a great
deal of uncertainty, but it was duly rewarded on the
third day when a meal was taken with such keenness
of hunger, such unction of relish as recalled the early
days of girlhood.

Now in the course of two or three years every local
trouble disappeared, and with it not less than forty
pounds of weight that was abnormal, and that had
been borne for years as dead weight. It required only
a few weeks, however, to get a most surprising degree
of relief, and there has been added a new lease of life
all the brighter because of that cheer that can only
come from the rich blood of vigorous digestion. During the first few weeks, she often confined herself to

one daily meal, and she never again made the mistake of eating without hunger.

In this case there have been fully seven or eight years of absolutely thorough testing of this science of nutrition in the culture of health, and therefore the cure of general and local disease achieved and maintained.

There came another case of still more striking character in a lady in her seventy-third year. She was in a general condition of bloat that had been on the increase for some months. She had an annoying cough, and had become so deaf within a short time that she had become unable to hear the church bells, a great loss to her, and she was entirely unable to go upstairs without assistance because of shortness of breath. Belonging to the highest circles, the center of a large family, and of a large circle of relatives, a heavenly endowed woman in all that dignifies, elevates and adorns, physically, mentally and morally, her case invited all the science I could master.

I found her taking a late light breakfast, a more substantial meal at midday, generally within three hours, a light tea and always an apple before retiring, hence *four daily meals.* These were cut down to one cup of coffee earlier in the morning, and then a dinner at the usual time, a light evening meal and no apple at night.

The improvement began at once. In a few weeks the cough and deafness disappeared, the heart gained in power and a decline in weight went on. During the following summer, she became able to walk very easily about the city for one of her age, and one trip nearly a half a mile up a decided grade. She has often told me since, that, except her age she, has never felt in a general way any better in all her life, and this has been going on for four years. And now, in her seventy-eighth year, there is a mind as clear as crystal ; and she is an enthusiast over the means whereby her life was saved and enlarged to a screner and more comfortable old age.

Let me here tell you something in a slow-impressive-sort-of-a-way, with a hyphen between each word—let me tell you that when I got into a fully realized sense of what this treatment was capable of doing, to relieve and often to cure local diseases that had baffled trained specialists, my satisfaction reached its highest professional possibility, and I began to wish for an honorary degree of higher significance than the mere M.D., and affix of C. H. C., "Councillor in Health Culture."

With the revolution that went on in the physics of health culture, there began to be noticeable one also in the mental forces behind them. I only observed for a long time a general quickening of the mental forces as I had done of the general physical improvements. In my own case I only realized a higher average sense of cheer and force, without any thought of what might be going on in special faculties. As usual my attention was arrested by a case that was to call to my consideration special results in a very striking manner.

A student called upon me with disappointment deeply engraved on every line of his countenance, for the express purpose of having me thoroughly examine his case, and then to give him a statement that his health would permit no more schooling in the class-room. When a mature young man he had started from the depths of a coal mine to make a minister of himself, via Alleghany College, Meadville, Pa. Now the learned and able president, the learned and able professor, very much more learned in the languages of the dead than the languages of the living body ; very

much more learned in international law than the laws
of life ; very much more learned in political than in
physiological economy, saw no incongruity in this
young man's appearance in the class-room every
morning with his stomach full of beefsteak, with its
usual accompaniments, who was not to work it off
by wielding a pick and a shovel during all the forenoon. You see, from the president down to the
adjunct, there was not the slightest hint of the
" grab-bag " contest that must go on between the
forces in the stomach and the forces in the brain for
the limited supply of nerve force that ought not and
should not be divided. Even at best the stomach
would fall far short of its need, by reason of an engulfed
overload and its destitution of teeth. In the absence of
this vital knowledge that was not to be found in text-
books, he had managed to push his way along with pick-
and-shovel diet methods, until he was ready to enter the
junior class. By consummate application and perseverance he had been able to keep along in his classes and
preach every Sabbath at country stations ; but by virtue
of daily abuse of life powers that were constitutionally
weak, the inevitable time, the long invited, came. He
had been unable to maintain his class-room standing.
Said he, " I have become unable to fix the points
of my lessons closely in my mind ; and even when
I think I have them secure they take their depart-
ure before I reach the class-room, and my pride will
not permit this torturing experience to longer go
on."

I gave the young man a lecture, in which I unfolded
all the physiological science behind mental science I
then had, and turned him about and sent him back to
his studies, and it was not more than two or three
weeks before there was a happier face, and such an
enlargement of grasping and retentive power that he

was able to meet the best of his fellows in the class-room without subjecting his pride to crucifixion.

He had no occasion for a second lecture, and two years later he graduated, not only with honor, but with twelve pounds more of muscle than when he made his call for a certificate of class-room disability.

Another case came to me from the college, of a younger man, the picture of robust health, who found in a few weeks after his entrance upon his studies that his frequent attacks of headache that had been an experience of years, were becoming aggravated, and that also there was a growing inability to master his text-book tasks. Now this young man had given several of the "faculty" of a great city an abundant opportunity to exercise their skill on these same head-aches, but all to no purpose.

It was another beefsteak case, and he was, as the other, advised to give the brain all the mental force that could be mastered during text-book and class-room exercise. It was only a short time before he could enter the class-room fairly alive with reserve power for the duties to be met, and the headaches became only a disagreeable memory of bygone years. He became greatly enthused over his new-found acquisition, and he would take unusual pains to make known his appreciation at our chance meetings on the streets. "Doctor," he once gushingly said, "I would not take any price in money that could be named for my knowledge of how to preserve my life, if it were to compel me to go back to the dark days of ignorance and headaches."

I also had my attention called to the effects on the moral powers, by a case striking in its clearness of demonstration. A young man with a fine family of blooming daughters growing up around him, who needed to see in him a high type of manhood, was most terribly afflicted by a general condition of malaise that kept his temper in a state of such chronic irritability as rarely failed to drive cheer out of the home whenever he was in it.

He was not aware of any digestive weakness, but behind these symptoms were the morning steaks and the between-meal nibblings—he was a grocer with a retail liquor annex—the habitual daily stimulations, and their excesses.

This wreckage had gone on with its impairment of memory, and ability to conduct his business affairs with any ease or wisdom ; and so did he become more and more a cheer-depressing force to the lives of the amiable wife and the unusually interesting group of daughters. He met the better way of life in his most pressing need —he became a total abstainer, and forthwith there was a vigorous awakening of physical, mental and *moral* power.

The memory had a most striking gain as well as the physical energy ; the gain being all the more rapid because of a good constitution ; the absence of general and local disease, and the presence of a stomach that could be easily turned into power. I listened to this short speech. `` Doctor, you don't know how much better I enjoy my family than I ever did before ; they all seem so nice to me now ! Why, it used to be so with me that I would fly into a passion if even the smallest child would fall down and thus harass my irritable temper with its outcries.''

Was there not eloquence for me in such language, when the expression was surcharged with meaning as if every syllable of every word were the language of the soul, regenerated, purified ?

Ah, gentlemen and ladies of the medical profession, you are getting no such results by treating the symptoms arising from perpetually overtaxed stomachs, with a carefully assorted array of remedies, whether these remedies be coarse or whether they be refined !

LECTURE XV.

EVOLUTION OF DISEASE.

SUPERFLUOUS FLESH—APOPLEXY AND SOFTENING OF THE BRAIN CONSIDERED.

My Friends the Readers :—

I have enlarged my audience this morning by an invitation to a few personal friends who are heavily handicapped with overcoats of fat, to be present to hear my views as to the origin and development of disease.

When my attention was called to the very clearly manifest improvement in local diseases without a resort to any local treatments, very naturally I would want to know how this could be. I had, as I told you, solved the question of the loss of weight in dropsical conditions early in the history of my studies, to my own satisfaction.

I appear before you this morning heavily weighted with the importance of my theme, and all the more because of the difficulty I shall have in making my conceptions clear to your lay minds. What more important subject than the origin and development of disease, if it can be made clear that its development is very largely a matter of improvidence of the individual, and not of the providence of God ?

It may be assumed as a fact that every human being is born with a tendency to some disease through heredity, a tendency, as I have told you, that fixes tho natural limit of life.   This tendency or constitutional condition determines whether death shall naturally come within one hundred minutes, or one hundred years. To give a more definite conception to what may exist in constitutional tendency, I shall embody the idea in the expression, structural weaknesses, local or general, due to heredity.

It is my conception, then, that one person is born with a structural weakness of the nasal mucous membrane that is to become developed into a catarrh through avoidable evolution.   Another person is born with structural weakness of the throat and bronchial tubes that is to become developed into catarrhs associated with the annoying hemming and coughing efforts to relieve.   Another is born with structural weakness of the ear-passages, that, duly developed, is to impair or destroy their functional power ; and so also of the eye, and you of the *heavy overcoats*, it is my conception that some of you have been born with structural weakness of the blood vessels that circulate to the utmost cell of the very centers of vital power, only awaiting a due degree, perhaps, of largely avoidable development, when you will go *down like beeves beneath the stroke of the ax* ! !

You may well be startled by this statement, and I will tell you in advance that I am going to do my best to make you believe that there is a great deal more truth behind it than will make it a cheering subject to contemplate.

What do I mean by structural weakness due to heredity ?   I mean just this : The vessel walls are thinner, the contractile fiber is smaller and weaker, and all the structure of the parts has less tone than have parts not affected through heredity.

The gravity of this condition is always a matter of degree originally. You may understand, then, that it is my conception that these weaknesses are constitutional, enduring, and hence always a menace, according to their gravity, to human comfort or to human life ; and that they accurately gauge the constitutional, the life-sustaining force of the individual ; that by no human means can they be raised above the hereditary design in strength.

Normal health may be defined as that condition of the body in which the digestive machinery is able to respond to every need arising from destroyed tissue, through mental or physical labor.

The first step in disease, then, is the first loss of balance through whatever has impaired this machine-power, and hence the parts structurally affected are the first to feel the loss. It is my conception that with this first loss of balance, the contractile fiber, weak through heredity, begins at once to lose, through lessened nutrition ; this permits a gradual dilation. Now you were told how dilatation, distension of the capillaries and blood-vessels of the stomach, was incited by the irritation of the alcoholics, and the functional power of that organ was thereby diminished by subjecting all the intercapillary intervascular structures, including the gastric glands, to a life-depressing, strangling pressure. How could this be otherwise when the finely meshed capillaries begin to enlarge through the force of irritation, or through the slower process of passive dilation, from defective nutrition, in other parts of the body ?

Now to illustrate by a case. A little daughter and an only child is born with a structural weakness of the bronchial tubes. By a due course of lowered nutrition, by reason of a continually overtaxed

stomach, that began at birth, and was persistently
continued, these parts became subject to an evolution
tending to disease that did not become manifest in the
usual symptoms, until the third year, when there
began to be a hemming that went on unnoticed for a
long time. Then came the coughing spells that hung
on persistently with each cold, then began to be noticed
more or less coughing in the absence of colds, that
excited apprehension. Vigorous treatments with
home remedies were applied ; then the various
advertised specifics that, while in use would suppress,
but failed to cure the cough. The years of anxiety
went on, and the winters of aggravated disease had
their nights of broken rest for the child ; of broken rest
for the mother, attended with the paralyzing apprehension to her of the gravity of the disease. All this
was endured up to the eleventh year, and yet danger
seemed as far off as ever. Her case had been subjected
to the skill of several of the most experienced and
learned of the medical fraternity, but it had proved too
obstinate a case for the science of remedial therapeutics.

Now what was the condition of things at the seat
of the disease. With the first loss of balance of
nutrition that began with the first untimely meal the
blood lost a little of its richness and a **tax** was laid
upon vital power. The little blood-vessels began to
dilate into pouches, hence subjecting the intervascular
structure to the strangling pressure. The blood
would circulate much slower in these pouches, and
hence, by reason of its abnormal thinness, there would
be a tendency of some of its water to escape more easily
through the thin walls, to become thickened with the
natural secretion of the part, and so form a discharge
that was behind the irritating cough. Now let me
string these conditions into line.

1. Thinner blood by reason of bad eating habits.

2. Dilatation of the weak vessels from lessened tone of the contractile fiber.

3. Intervascular pressure to squeeze the life out of the tissues.

4. Slow circulation through the dilated pouches.

5. Hence favoring the escape of the water from the blood to aid in the formation of the sputa.

There is no germ theory in this conception of the origin and development of disease. Such is my conception also of the origin and development of throat and nasal catarrh.

Now, readers, you must easily see 1. that a return to the normal condition must be by a reversed order of evolution, and 2. that this cannot be done through sprays and douches. The blood must first be made richer, before those dilated pouches can be reduced to the original constitutional size and so relieve intervascular pressure ; and the way this is to be done you have already been made aware of.

Now the little girl was at once put on to a light breakfast and the heavier second meal, etc., with all between-meal nibblings prohibited, the mother being assured that the cough should be absolutely relieved if she duly guarded the stomach from all abuse.

The result was an evolution in *reverse.* As there have been three winters since with scarcely so much as a cold, and the cough entirely relieved ; and, more, with greatly increased general health, the mother feels that a life has been saved.

How am I to account for this cure on any other theory ? We can never tell whether our science is advancing or whether we have any science unless we have illustrative evidence.

A man called upon me not only because of a bad case of nasal catarrh, but because of a deeper seated trouble, an irritation in the depths of one of his lungs. With a brother recently dead from consumption, he was naturally full of apprehension. He had been duly informed by an able specialist that there was a possibility of trouble of a severe character. The call upon me was in early autumn, and as his affairs did not prevent, he was advised to take his first meal at 10 A. M., and the next at 5 or 6 P. M. The evolution began at once. At about the first of the following January he began to follow his usual avocation, that of a carpenter, and became able to perform out-door work even in very cold weather with no food in his stomach during the entire forenoon, and with an ease, cheer and energy, that confounded his associates. The catarrh and lung trouble had rapidly declined through evolution, and there was added in time twelve pounds of muscle.

Now there was a daughter in the house who had an acute winter nasal catarrh, that very nearly prevented nasal breathing during nights—a condition that most mothers know all about. Of their own volition the parents had begun to confine this, and a younger child to the two meals daily, which soon resulted in a loss of all desire for between-meal nibbling and lunching. I saw these children after a few months of such living, and as they were the only children in the entire city that were ever put on such a regimen, you will be interested to know the results. In the oldest, *absolute relief from the catarrh*, and in both such development of the general health as to give their faces an ivory-like polish, and a delicacy of tint from Nature's own rouge, such as I had never seen before upon the faces of children.

Never before had children been kept on two daily

meals, perhaps in the entire state, and perhaps never were children kept from all uncomfortable sense between meals, of hunger or cramps ; or who had kept up a higher daily average in amount of food eaten, for the health was so steadily maintained that no meals were ever missed.

Let us go back and consider a little as to what heredity does for us. It determines everything except what comes from culture. It gives us all our mental faculties and determines the force of each ; it gives us all our tastes and determines whether they are to move us along the paths of virtue or vice, and it determines what the exact physical form is to be, even to the most delicate outline ; so does it determine what the exact weight is to be ; and let me assure you there can be no cutting down of excess of weight through remedial means without cutting *power* down with it.

Now you of heavy overcoats and vast rotundity of belt, I address myself to you in particular, for I am now to talk to you about a disease before which physicians stand, bent in lowest submission, as powerless to avert as the lightning's stroke, and no less powerless to relieve,—a disease, "a mysterious something that gazes like the eagle and strikes like the thunderbolt."

"*Young man, keep a clean record.*" And after this his greatest human admonition, down went John B. Gough, the eloquent lips closed forever.

A few years ago a great man, great in a splendid physical, mental, and moral personality, a Secretary of the U. S. Treasury, delivered a great address with all the ease of vast powers in reserve, with every line of expression beaming with living light and force, before a New York audience, great in its intellectual culture and power in the affairs of the nation. The

13

address reached its easy close, and before he could even enjoy a single relishing congratulation, down he went with a crash.

A New York magnate, great in the millions under his control, received another magnate, great in the number of iron horses under his sway, in the quiet parlor of his palatial home, and while vast interests were being discussed an arrow from an unseen bow reached the center of life and in an instant the vital spark was extinguished. And so all over the land men are dropping who seem to be in the prime maturity of their powers, and too often the most estimable of citizens. But you gentlemen of the lean, wiry build, you too go down, but you get up again to drag about one half of your bodies over your slow ways only to anticipate the inevitable second attack, and if you arise from this there is little left of your life that is not a tax to your friends.

These "strokes" are fearful to contemplate—only the other day one of the greatest of American surgeons went down in the maturity of his powers when there seemed reserve enough in him for a generation of work for his deft fingers, dictated by the master brain. And so went down to death "the greatest preacher since Paul preached on Mars Hill," while yet at the entrance upon a green old age with reserves, infinite to draw upon. Did that reporter know whereof he wrote when he said, that *baked clams* were a part of his evening meal? Undigested clams in the stomach when that sleep was entered upon that was to know no waking!

Gentlemen, it is a most fearful thing to have a structural weakness in those pipes that convey the "elixir of life" to the nerve centers of all physical, mental and moral force! And yet it may be considered a structural weakness due to heredity, entirely, primarily. What a boon it would be to humanity if some scheme could be evolved whereby *development might be avoided !*

In the consideration of this disease I can give you no illustrative cases. I may tell you that death from apoplexy is due to a bursting of brittle arteries in the center of life, that the bursting is due to a fatty degeneration of the coats of the vessels in which, not only fat, but lime also becomes a foreign element. But the revelation of this fact makes it seem all the more ghastly ; it gives you no relieving light.

You gentlemen who are to go down only once are, as a general fact, among the very best citizens. Your minds are well balanced ; and there is such manifest power in your mental and physical make, such all-abounding cheer beaming from your faces, that your very presence has a soothing, subduing effect, upon the nervous, restless activity of us lesser weights, of whom I myself am a striking illustration. As we see you walking your easy cheerful ways through this world, or occupying your easy chairs in the most healthful repose, we are impressed with the lesson that the world can get on for a time without our bustling services, while we may indulge an enjoyable rest.

In your structural make, in your moral and mental qualities, you are our ideals in man's largest estate. You all enjoy life ; you go to your well loaded tables with keenness of appetite and indulge without let or hindrance, for your stomachs will kindly expand through hereditary power, and will rarely admonish you of an excessive tax laid, or deposited in them. You go on your easy ways, fairly surfeited with reserve power until you near that period of life when ambition has reached its highest tide, and when a desire for "new worlds to conquer" has greatly declined. For this reason your incentive to work off the accumulations of table deposition has suffered a marked decline, hence you actually take less exercise habitually than before, a fact that becomes more and more marked with advancing years. But your disposition to exercise has now become subject to another depressing force of a very marked character; you are beginning "to take on flesh;" your overcoats have begun to become heavy as if they were becoming padded with a slow addition of shot.

You cannot exercise as you did before without getting out of breath, but you still feel just as well, when in repose—but you regret exceedingly your loss of physical power. The years go on with increasing inability to exercise without discomfort, and the overcoats become more and more thickly padded; the rotundity of belt continually enlarges, and finally a time is reached when some little excitement shall send an extra tension upon those brittle pipes and *down you go.*

Or another condition of things may be incited in the life-centers. The arterial coats may become so thickened that they can no more act as channels, and hence large areas of brain substance become subject to a diminishing supply of the elixir of life, hence softening of its structure and decline in mental power that may reach imbecility, when you become a tax to others.

How inscrutable, how mysterious are the ways of "Divine Providence" in this matter of the origin and the development of disease!

In my younger days I saw an eminent doctor of divinity go slowly down to death by reason of occluded arteries, in the center of his life. With a splendid form, with a head and face ideal in classic outline, there was only mind enough left to utter the most imbecilic ideas in classical expression; all else had gone, not even the sense of the need or the propriety of clothing.

In the earlier days of my practice I used to often meet a splendid personality daily upon the streets of my city. His form was elegant in all its outlines, and, with the serenest of mental temperaments, admirably balanced to enjoy life, he was the happiest of men in appearance as he actually was in fact. He was an ideal man in all his family and social relations ; and well he could be, for behind all his mental and physical needs was a stomach capable of any amount of taxing without complaint—and why not be cheerful with the sense of physical comfort always perfect ?

He too began to take on flesh as he reached the high tide of his life, and from thence on as he took less and less exercise, there began a slow decline in mental power that ultimately reached an almost disabling degree, while still in the prime of a later manhood, for any important professional duties ; and but for the supervention of an attack of acute disease, imbecility or a "stroke" would have been the sequel.

How mysterious, how beyond all finding out are the ways of "Divine Providence" in the origin and development of disease of such withering, blasting power over the very center of life itself !

The materia medica can little avail here, for why go to the physician when health seems so perfect : "the sick need the physician," and you are so abidingly well that you never meet him without a suggestion that you are only too happy that you need none of his wares.

Can there be any preventive measure that would

seem to be of any avail against the bloody, withering hand of apoplexy ?

Readers, I will meet you again on this subject, and you of heavy weight shall occupy the front seats, and you lighter weights who may think there is a bare possibility of going down, later on, to arise again with maimed bodies, shall sit next to the front.

LECTURE XVI.

EVOLUTION OF DISEASE.

My Friends the Readers :—

A few years ago England's greatest preacher went down to a slow death. As he approached the mid-period of life, the "flesh" began to accrete, pound after pound; but his was an eccentric stomach, and hence every meal was a burden of selection, of exclusion. There was an abiding torture of uncertainty as to what foods must or must not be eaten. And so those later years of his great life were burdened with those ills due to a stomach that never had rest, even during the vacations at Mentone.

Had those meals been duly separated until the axman's appetite came, there would have been a close approach to the axman's power of digestion, and the axman's luxurious indifference as to what food to be eaten, if it only met the relishing sense; and between the eating scenes, no matter how widely spaced, there would have been in lack of mental force because of the ample, the undue supply, the burdensome supply, of brain food stored away which was far in advance of any possible need. And so died the great preacher, years before his time.

At that far-off home by the ocean where these lectures were reduced to form, there died a few years ago a former fellow-citizen and schoolmate, a lawyer of great ability and profound research ; died just at his entrance upon the age of his primest maturity, after a lingering approach to a final sudden death. He too had a stomach, but medical science never hinted to him the need to treat with due consideration a most willing, a most obedient servant.

What would it not be worth to humanity could only some scheme be devised whereby the bloody stroke of apoplexy might be stayed or the withering, blightening grasp of the unseen hand upon the very power-center of life might be relaxed !

How is it that the walls of those life-pipes in the center of life become changed into a crude mixture of lime and fat with the contractile fibers, the life-guards, powerless to save because slow annihilation had gone on with them ? Has there been evolution in all this ?

The first step in every disease is the very first moment when digestive balance is lost. *With each one of you that evil work began with your very first meal.* This meal was taken, forced upon you by the nurse, before Nature made any demand ; in due time trouble began, every outcry was interpreted as a signal of hunger, and the oftener you were fed the oftener the solemn stillness of all the air was broken by your nerve-lacerating music. Your meals all through the first year of your life were regulated by the tunes of crying.

You who were born with good stomachs were the good babies, and you born with the weak were the bad ones. The hapless victims of a heathenish, a barbarous code of dietary ethics ; Nature had her revenge by nights turned into days with their taxing cares innu-

merable, all abounding, crucifying. You got through
your first year by a most heavy involvement of parental endurance. You got through the first year as
marked illustrations of the doctrine of the survival of
the fittest, but your weaker brothers and sisters fell
by the wayside.

> " God moves in a mysterious way,
>   His chast'nings to perform ;
>   He plants His footsteps in the sea,
>   And rides upon the storm."

The sympathetic pastor's enlargement upon the mysterious, the inscrutable ways of " Divine Providence,"
that little hinted of a more refined way of tossing the
innocents into the open mouths of the river Ganges !!
And the bereaved were impressively enjoined to submission.

Why language so strong ? Because—because it has
the strength of truth in every syllable. But how do
you know ? I will tell you how I know. I know by
the most ample experience from actual tests that the
highest health comes from the rigid exactness in the
feeding times ; with only three feedings in twenty-four
hours, and with the result of reaching the most exacting demands for growth. I know from actual tests that
on such feedings there is the most complete absence of
any hunger uneasiness until the feeding time is actually
reached, and the signal is not a wail of anguish ; and
at night, without any feeding, there is such perfect
sleep as only a perfectly nourished infant reaches, and
there is also parental sleep. I know from the most
ample of tests that four meals in twenty-four hours
are all-abundant for every demand of rapid growth,
and that by their more perfect digestion they reach
a higher sustaining power than can be realized from
five daily feedings. Behind functionally weak diges-

tion is a structurally weak stomach ; five daily feedings mean for very many lives an habitual overtax, and by as much a disease-cultivating process.

I know by the most ample tests that by such regular and limited feedings there results that highest possible health average which makes the infant a joy, a solace, and not a health-destroying, happiness-blasting care. And I have a right to assume that this high average in richness of blood and tone of the vessel-walls by as much hinders the developing of the possible catarrhs, nasal and bronchial, and every other local disease that man is heir to ; and no less the possibility of even general diseases, not excluding consumption with its germs of origin.

Life went on with you. You with the strong stomachs had the best time of it ; you with the weak, had your frequent colds, your billious attacks, your days and weeks of not eating, during acute sickness, only as forced upon you by the supposed need. Your various local diseases became more and more marked under the stimulating power of your daily irregular feedings.

And now mark this, *you with the strong stomachs, from the days when you began to do your own feeding, you habitually ate more food on the average in proportion to the exercise taken, far more, than has been the case with you thin, restless, nervous mortals to whom an easy-chair is an abomination, and who never can keep still one blessed minute except when enfolded in the arms of Morpheus !* Do you think there is any significance in this remark, any slight hint to be drawn from it of a possible excess of lime and fat taken into the system, and a very slight evolution incited and continued *in the very center of life's power ?*

You went on until you reached life's mid-period, and you began to take less exercise than before ; but you with the strong stomachs very much less because you became more oppressed by increasing weight. But *you* went on eating, eating, eating just the same, the very, same. There is no thought of cutting down to meet the actual demands of expenditure. And so you added, by this eating in excess, "not strength unto strength," but pound unto pound of *sheer dead weight*, that as it went on made the easy-chair more and more a luxury. The heart becomes weak, and so you must walk slowly. Perhaps your inherited tendency is the gout, and so your existence has become a torture, as also the existence of all who administer to your wants, wants beyond enumeration, beyond the remotest possibility of pleasing, even with hands angelic.

Every day of your lives, at each meal, there was more lime and fat getting into your systems than the vital forces, the vital machinery, had the remotest need of. And what becomes of it ? Why of course you easily see and oppressively realize what becomes of most of it, *but—but—some of it—will be found in the degenerate walls of those life-pipes, post-mortem, after the clots are cleared away.*

I simply don't know that the degeneration is the outcome of years of evolution through eating habitually in excess, as I cannot prove it by evolution in reverse order. But, in hundreds of cases, of which I have given you a few striking examples, I have fully satisfied my mind that the evolution of disease is from a structural weakness, due to heredity, as the primary condition, and is always a matter of development through malnutrition in which *avoidable cause* is by far the largest factor.

I address myself to you, heavily-oppressed readers, now in particular. You are heavily borne down as if in the relentless grasp of an "old man of the sea." And how are you to loosen his grasp?

If you think the theory is plausible you may feel very much disposed to begin to unload, not only as a matter of greater personal comfort and strength to be realized, but also because of a lively sense of personal danger involved in this matter of an evolution of disease through avoidable cause, which you begin to fear is going on within you.

And how is this to be accomplished? You must definitely understand, as a first proposition, that you have accumulated a supply of reserve, predigested food that should keep your vital machinery in motion for from sixty to ninety days, even without the digestion of a single mouthful of food during the time; and this without involving danger to life. You are not going to be attacked by disease after the brain gets down to drawing its supplies from the tissues, with the stomach in repose.

As a second proposition, you will naturally need to reverse the order of accumulation. As you have been eating far in excess of the demand arising from your daily exercise, so must you reverse the order, the habit. By eating less than the demand you will compel the brain to draw upon the reserves, and this is the only way that you can lose your weight without the use of means that are destructive to health; it is the only way that will add to the strength of the body while the decline in weight is going on. You noticed the relative loss of the muscular system as revealed in the "Mable," now you are going to avoid this loss by a due amount of daily exercise. You will limit the loss to the fatty tissues alone; there will be some loss in weight through the absorption of water that gets into the tissues more or less in all cases of low nutrition.

But how are you to take your meals so as to get this evolution into satisfactory progress ? That will depend on how anxious and persevering you will be. I once advised a lady heavily burdened to eliminate one meal per day, and as results soon became very appreciable, and getting in a hurry, she eliminated a second meal, and was not disabled thereby from her household's heaviest duties ; and she succeeded in getting a reduction of thirty pounds during one year.

It is a process that you can push safely well on to the skeleton condition, if you cut the food down. You can determine in advance just what loss of weight is to be reached ; for it is simply a problem of endurance and mathematics. If you cut your food down by one meal per day, and take a due amount of exercise, a certain decline will follow. If you exclude the second, a greater result will follow, and still greater if you get down to one half a meal a day ; the loss must come just as it does in a case of fever. The brain requires its needed amount and the fat seems to meet the demand in a very large measure, and this will disappear inevitably when no food is being taken into the stomach.

A lady in the country weighing two hundred and sixteen pounds, a very hard worker who had a variety of ailings in addition to her excess of weight, was persuaded that she could well afford to stop all her affairs in mid-forenoon, and take the pains to get as good a meal as Nature could order, even if involving the unsocial way of eating alone, and to make this meal suffice until she could take the second with the family in the evening. This plan she adopted and rigidly followed, and there resulted within two years a loss of fifty pounds, and a most marked gain in the general health and local ailings.

It often takes a long time to cut the weight down to a satisfactory degree, but the degree reached is always a question of radical measures enforced and continued.

A supreme advantage of the method is realized in the manifest return towards the vigor of earlier years. And I think without any question in more vigorous condition of the vascular system generally, and perhaps in that *center of centers* where *tone* is so much needed. Why not its *prevention* if there is an *inherited possibility ?* Why not relief if the evil work has begun ? And why not take it for granted that this theory represents a fact in the evolution of a condition that makes a stroke of apoplexy a fearful possibility, a *constant and most dangerous menace to your lives ?*

Let me state this theory as a *fact, self-evident, self-conclusive* in another form of expression. *From the time the very first meal went into your stomachs, even before you yourselves had made any call for them, then began a change in the life centers, infinitely slight in character, perhaps beyond the power of the microscope to reveal, but nevertheless a change that was to slowly develop in due harmony with violation of digestive conditions—until at last there is a degree of degeneration of the life-pipes, such rotting of their walls, as only to require a breath of excitement when the floods burst through, and then there began— an evolution of decomposition ! ! !*

It is an anomaly in the development of disease that there can go on such blastings, such blighting, such destructive work in the very core of vital force, and give no hint to the victims of the fate that is in develop-

ment for them.   Other diseases warn in time, but this
never.

Now as you believe without any question that the
various lesser ills and diseases are the evolutions arising from disordered digestive functions, and that there
can be no return to the normal condition, except
through evolution in reverse, you must see that your
safety involves a like reversal of the evolutionary process.   Now, year after year, you have been adding
to your weight by accretion, so must it decrease by
absorption.

My oppressed friends, I have no language to express
my sense of the importance of this theory to you as to
the origin and evolution of a disease, which is so in-
sidious as to give not the slightest hint of its existence
until all avertive means are too late.   If it be based
on a pathological law, or on physiological law under
the action of disease, I have no words with which I
can express my convictions of your need to begin at
once to get yourselves in line with the laws of your
lives, in line with "constitutional law" that will result in an immediate, manifest improvement all along
the general lines of health ; in all abnormal local conditions, that must include improvement in a most dangerous disease that perhaps may be yours, unless it be
excluded by an anomaly in nutrition in the cure of
disease.   I certainly do not know that I have ever succeeded in inducing a reversed evolution in this disease ;
but cases where there have been reversals in the order
of the accretion of fatty tissue have been striking and
numerous.

My heavily-laden friends, there are no victims of
disease to whom I address myself with such keenness
of interest and with such force of conviction as I do to
yourselves, for your enemy is in ambush ; he is ever

engaged upon a "still hunt" after your lives ; ever ready to give his fatal spring to bear you down to death or to fasten his relentless arms upon you as the "old man of the sea," whose heavy weight must ever be borne along your weary ways, until at last the "tired heart shall cease to beat," and you enter upon your final, your eternal rest.

In order that you may get a still clearer conception of the science of cure, of the science of nutrition, I will give you a physical demonstration that may almost be seen with the naked eye.

When my friend the publisher met this "New Gospel of Health," as he is pleased to call it, he had been in low health for some years, and as I was enlightening his mind from time to time as to the science involved, through correspondence, I called his attention to the fact that through the toning up of the entire vascular system, the little microscopic capillaries, arteries and veins in the face which, by dilatation, had thrown up the skin into folds, and hence causing a coarse, soft, flabby condition, would begin to so tone down in size, as to make the skin not only smoother but more dense, and therefore if he were shaving himself, he had already found that his razor was cutting better than before, and because the beard would be held in firmer grasp, and hence easier to cut, would be softer as well.

Now it so happened that before entering upon this new system of living, he had been compelled to take his razor to a barber to be put in cutting order ; but it was of little avail, and he was about to return it for still further effort, when the matter got delayed by reason of attention engaged in the new field of interest, until he became aroused over the mystery that his razor seemed to be sharpening itself right along ! And great was his surprise, keen was his interest when he found that not the cutting edge but the skin had been undergoing a change !

When I went to that home by the sea to reduce these addresses to form, worn, much exhausted by a summer of unusual heat and an unusual succession of life-or-death cases that went on for months, my razor required deft handling, and the experience was not void of discomfort, and if, by a chance, one of the little vessels would get nicked, there would be a protracted oozing. Now, under the revival of a change of air and unbroken rest, new force, new tone came to the utmost cell, and that razor seemed to get a keener edge at each using, and the nicking more rare and the oozing of short duration—all from better tone.

Let us take a face of early mature age, sallow, of greasy complexion, and, for instance, with a hint of furrows ; there are studded over it pimples great and small with occasional patches of dark red congestion, a form of skin disease, and of long-standing, in fact it has stood repeated efforts to remove by the various washes and healing salves. Behind this condition of things has been enforced or indifferent eating, for months and fitfully during life.

The new order of living is instituted, and with the first relished meal a thrill of electric energy is sent through the capillaries. At once the pimples begin to grow smaller and the ugly-looking, defacing patch begins to lose its dark color and to contract its boundaries, and all over the face renovation is going on. The clearing-up process goes on day after day, the smallest pimples have been fading away, and the largest have now become the small ones. The patch is still smaller and the color is much brighter. The furrows, the wrinkles have also become less marked, and

14

the minute tracings, wrinkles in miniature, which the
magnifier would have revealed, have totally disappeared.

Evolution goes on and at last there is a skin free
from all abnormal defacings. The capillaries and the
minutest arteries and veins have become toned down
to the normal size with the skin becoming more and
more dense, compact in texture. The little arteries
are nearer the surface now, and hence the skin shows a
diffused, delicate shading of Nature's own rouge, that
compares with the diffused redness ordinarily seen
upon faces as the most delicate tints of the artist compares with the coarser colors of the house-painter.
And the skin itself, why, it has to be studied for a
time with the eye directed to the science of nutrition,
there must be eye culture before its *plush-like softness*
and delicacy of tracery can be fully appreciated with
artistic vision ; but the most ordinary eye can see
enough to call out expressions such as "Why, how
much better you are looking," and the place of all places
to look better is the face !

While this evolution goes on in the face another is
going on in the eyes. Those delicate, transparent,
nutrient vessels that supply the retina and cornea with
daily food become toned, and there is greater reflective
power in the mirror ; there is a clearer power, denser
texture in the front window, and hence a finer delicacy
of vision whereby the beautiful in nature is seen as
never before.

And while this is going on, the structures entering
into the formation of the sense of taste are alive with
unwonted energy. Little by little the sense of relish
develops in acuteness, and little by little there is a
revolting against the acrid juices from the pipe, the
Havana and the " cud " if you habitually use it ; and

still behind all this is that higher sense of physical
comfort, that absence of uneasiness, that indescribable
restlessness that calls for the anæsthetic, "peace be
still," smoke or chew. Tobacco is an anæsthetic in
degree. It always answers for your between-meal
lunchings ; it has a subduing effect upon the complainings of Nature for your habitual violation of her laws ;
you smoke before your meals that you may await
them with more composure ; you smoke after meals to
allay the murmurs of overtaxed digestive processes ;
you are always in a more or less uncomfortable state
after the transient composure of the anæsthesia has
gone. And all this artificial comfort is at the expense
of that keenness of relish that is your right to have as
a pleasure, but also Nature's right to have that she
may select the class of viands that will yield the largest
per cent. of her actual needs. That infinitely important nerve, the gustatory nerve, was never constructed
to endure numerous daily baths in so acrid a fluid as
tobacco-juice without such a loss of functional power
as to lower energy-power in the utmost cell of the body.

But you say that very many users of the "noxious
weed" do live to a great age. Very true. At the
Infirmary where I was in attendance was a slim old lady
that sat erect in her bed and drew consolation several
times daily from her clay pipe—she was in her ninety-
seventh year. I recently met a man on one of the streets
of my native city in his ninety-fourth year who said to
me, "I feel perfectly well, eat and sleep well, the only
trouble I have is rheumatism in the limbs ;" and he uses
tobacco, and all his life has used alcoholics in moderate
quantities—but what does this prove but unusual
hereditary, or constitutional power, reinforced by an
easy and unusually serene temperament.

"A. Ward" knew a man who took one drink of

whisky every day and yet lived to be one hundred years old ! But he would not have young men infer thereby that by taking two drinks every day they might live to be two hundred years old. Only the fittest live to old age, and the very old die prematurely young, in fact, if they have habitually lived in violation of constitutional law.

My laden hearers, I thought I should be through with you in this lecture, but as I have gone on, new conceptions of your need have so pressed upon me that I shall have to call you before me once more. You will go to your homes in a more thoughtful mood than you came, and if you chance to awaken during the night, you may find it a little difficult to at once get back to the temporary oblivion of all worldly care or apprehension, for there may be the disturbing thought that avoidable wreckage has got far into the danger condition.

You will meet me again when I will give you all the cheer I think I may rightfully impart through in-ferential illustration.

LECTURE XVII.

My Heavily Laden Friends :—

The changes that are so noticeable in the skin, in vision and in the sense of taste, as the direct and generally very early result of increased nutrition, are also to be observed in the sense of smell ; this you would very naturally infer from the relief experienced in cases of nasal catarrh.

The change in the moral sense is often very striking as I have already called your attention to in one striking case. If you shall be induced to put yourselves under this natural diet regimen, if you are Christians, you will become a great deal better Christians ; if husbands, a great deal better husbands ; if wives, a great deal better wives ; if parents, you will be a great deal more kindly, considerate, reasonable in all your relations to your children. There will be such moral strength added that in all your relations to the family, to the church and society you will become *better, stronger men and women.*

I had a most striking illustration of the improvement that may be wrought in the disposition, by adding to the power of nutrition, in my own family. The last three years of my life, before I began the life-saving regimen, were the three first years of my youngest son. All through the first year he was fed on cow's milk on the canonical plan every two

213

hours, that most of the books enjoin, but which parents
very rarely carry out. In this case it was very rigidly
carried out. Except in regularity I was in the dense
darkness of heathenism so far as having any clear
conceptions as to the difference between getting foods,
liquid or solid, into the stomach, and having it duly
disposed of after getting there. The son seemed to be
a sturdy, well-nourished child, but he was habitually
encountering the same digestive overtaxing, was in
the same bloated condition and was subject to the same
bowel trouble as was the case with the young child I
told you about in a former lecture ; and it continued
through his first year, for there was no book for me,
by Dr. Page, to set me on the right track, and so
all through the first year I thought I had the worst
dispositioned child that was ever born to afflict his
parents.

And during the ensuing two years of his life, there
was such habitual irritability of temper as to keep
paternal affection down to the very lowest ebb. In
addition to his three regular meals he always had his
small between-meal lunches, and during the apple,
peach and pear season, because these are always
" healthy," he always had the privilege to improve
his health by such means to the fullest extent of his
desire.

In addition to this wreckage there was a damp cellar
that had never fully recovered from the washings of
several unusual floods, and, in spite of the best sanitary
efforts to put it in good condition, the house was always
below sanitary need.

As the evil work went on so was I kept in a state
of such chronic irritability of temper that I felt most
of the time that this son had come as an afflictive dispensation of divine Providence.

Afflictive dispensation of divine Providence ! It was a case of action and reaction, of cause and direct, immediate effect, of sin and the sinned against, and it was a case of " thou art the man."

In looking back to those darkest years of my life, to the suffering years of his life, with a fully realized sense that for both of us it was all due to avoidable cause, there comes a burning desire that I might go back and have another thirty years to practice a higher science in the healing art, that I might be endowed with apostolic powers of persuasion and go before all the people with an unfolding and uplifting of Nature ; that infancy might be saved from the barbarism of ignorance and from the insanity of irritability of temper : "Insanity," do you say ? Yes, beyond all question. It is such in degree, and it only needs the due aggravation and duration to wreck the strongest intellect that was ever inherited by man. There is a physical basis in all cases of insanity as there is always in that electrical, excitable condition of low estate, that only needs a spark to cause an explosion.

As this son entered his fourth year he was taken to a home freshly made to order, where sanitary conditions were nearly perfect. And there was at once a noticeable decline in the unloveable in his general demeanor ; and as he gained in physical strength, month after month, so did I cease to question this gift of Providence : and later on when I entered upon a higher physical life myself so did I consider this son's faulty, diseased manifestations with calmer mien and softer spirit, and, as the years went on, and there developed a disposition, through his higher physical living, that permitted a union of his soul with mine, unbroken by a harsh outgushing from either towards the other ; and I might well wonder that so great a happiness has been vouchsafed me through one so deeply sinned against.

Ah, readers, there is no exaggeration in all this, and such tragedies of emotional life are part of the experience of families in more or less degree everywhere, and moral force bends before them. The irritable are always physically sick, and sickness always demoralizes moral no less than physical and mental force.

Now, my laden friends, you are keenly interested in this matter of loss of weight, and all the more since you have become aroused to the possibility that your enemy, your possible enemy in ambush, may be reached and dislodged. You would like to feel that by better living habits you can have some assurances of living out the natural limit of your lives, and thereby have the privilege of finally ending those lives by natural deaths. You have begun already to catch the idea that improvements such as I have outlined must mean for you, by as much as you can develop them in your own persons also, an arrest of development of disease, if, perchance, it exists, and in that *central core* of all living force. Indeed, you can scarcely doubt this, and more, if the human face can be cleared of its pimples, large and small, little diseases in fact; if the defacing patches can all be cleared away, and the muddy, the sallow, the tallowy-appearing skin can be softened, condensed, implushed, tinted with Nature's matchless rouge, and enlivened with an outward expression of a higher soul-life within ; if the eyes above have begun to fairly glisten with Nature's own polish where was only dullness before, if all this can be seen to go on with the naked eye ; how can you doubt that a corresponding degree of relief must go on everywhere, even in the very center of the brain, and hence the ambushed enemy be compelled to relax his withering grasp?

Now let me tell you, as an illustration of Nature in disease, what resulted from a ten-minute speech to a man who had most pressing need to listen, and who also was endowed with a receptive mind.

His age was about forty-five, his weight two hundred and twenty five pounds, complexion sallow, expression cheerless. No local diseases known to exist, but bilious attacks very common. Habits— a steak breakfast always taken by reason of a supposed need and never with any relishing sense; in this temporized lunch pail it would be carried about until mid-forenoon, when exhaustion would indicate the ham-sandwich and beer. At noon there would be an enforced dinner, and in the afternoon another sandwich, followed by an evening meal very generally enforced. The more or less habitual feeling of discomfort engendered by such living habits, was a perpetual temptation for the anæsthesia of beer, and it was fully used for this purpose. And as the years had gone on so had the frequency of painfully acute attacks of digestive trouble become more and more frequent.

An evolution began with him, at once; and the accretion began to disappear; little by little the weight declined, and with it a decline in the desire for the relieving glass; and as the new life began to show its results all along the line, so did he become aware that the glass was having a less soothing effect than formerly, and so did his ability to say "no, thanks" with more and more emphasis to the *social* invitation to drink.

My old inimitable friend Artemus Ward has told us that his greatest speeches were all made before the bar, but that they were short; as, "I don't care if I do," and "No sugar in mine."

The greatest speech that can be made before such a bar is by one who, by right living, has become possessed with a regenerated brain which can enable him to say, *No*, not only from principle, but doubly *No*, because he feels not the slightest need.

The evolution with my laden friend went on ; his face cleared up and there was a ruddy change to be seen in every line. His step became elastic, the old attacks became lighter and further apart—in fact they never came, only, as he clearly saw, on special invitation.   His home became brighter, for the between-meal sandwiches were abandoned.   As the first meal, the morning meal, dwindled, so did the second or noon meal enlarge, and so was it prepared with a zest unknown in earlier years, because painstaking effort was always to be rewarded by a "well done, thou good and faithful wife."

In this case there was only one short speech made, but the   reply came some years later.   " Doctor, if I had not adopted this new method of living when I did, I don't believe I should have been alive to-day."

In order to still more deeply impress your minds I will draw another illustration.   Many years ago a tall, slim, large-boned young man seated himself in a professor's chair in a well-known college.   He was a man of great physical powers under the perfect subjugation of moral force of the first order.   He had all the force of will· to carry out to the utmost the duties that his cultured conscience laid before him.   In later years an eminent pupil once said of him, " 'Ought ' and ' ought not' were his most potent words."

He became the head of a great university, with Moral Science his specialty.   As years went on accretion began to weigh him down ; little by little, day by day, year after year, the intake exceeded the need as naturally regulated by demand and supply, and ultimately one of the greatest intellects his nation ever knew became all the more impressive, because a large physique became apparently, its adequate support. But in his last years there was a worn expression, an exhausted feeling, a tired brain. Was there a brain being starved by accretion in the life-channels by which the "elixir" was reaching the life-cells in a diminished supply? Perhaps not, for there was only the significant wearied feeling ; but in time came one of those sudden deaths through the brain, and a nation, a world, mourned over the loss of one of its greatest educators.

Had the law of demand been exactly heeded in the supply there could have been no such accumulation of vast stores of predigested food, the giant brain could never use. Why, gentlemen, watch a carpenter, for instance, as you see him up in the air handling heavy studs and joists, and rafters all the forenoon, and then think for one minute how much he will eat for his dinner, as incited by his need, and how much you will eat for yours after sitting in your professor's chair all the forenoon, in fixed air, more or less heavy with breaths perfumed with odors arising from steaks in a state of decomposition ! How will your dinners compare in the avoirdupois sense ? Is there any suggestion of habitual, shall I say *gluttony,* in the comparison ? "A word to the wise."

Now there chanced to come to my office a woman in her sixtieth year, tall, finely proportioned, not only physically, but mentally and morally as well. I had known her in her younger and slimmer days. Her cheeks seemed to hang in lifeless masses and there was little of color and life in the skin to indicate even the sense of depression that was a daily realized sense. She came to consult me for intolerable headaches she had been subject to since her thirteenth year, an ailment that had added more or less of torture during every month of those after years, and which much of the time had been a weekly irritation.

A call on me to cure a disease forty-seven years' standing ! And, after forty-seven years of failure, through all sorts of specific treatments, of remedies scientific, advertised and domestic ! In addition there was a spine injured, that would brook but little walking without a hint through headache. She was able to manage all the duties of her small family in a most uncomfortable way. I delivered to this lady a series of most emphatic statements as to what she must and must not do, without going into the science of reasons—only good food after the forenoon of fasting had created the appetite, and no drink but pure water.

I began very late in the history of this new movement to consider the question of tea and coffee. Had I begun an attack on these as an evil at an early period I should have raised doubts even in the minds of my attacked families, as to crankism that might disable me from the graver duties of my profession.

But here was a case of headache where the brain must be relieved from all stimulation. I sent my patient home somewhat enthused with the possibility that by her aid, her ancient enemy not only would be dislodged, but more, that her excess of weight would slowly decline, and, more than that, not only would there be this decline, but with it more and more energy to perform all her duties, heavy or light, more and more cheer in the performance of those duties, but still more yet, that there should come such relief to the injured spine as would permit very much longer walks, with very much less of complaints thereby from the injury. All of which was a great deal for one woman to conceive of as possible after so many years of suffering and failure of treatments.

Three years later she came again with an expression as it might have been had she just come down from some high mountain, fresh from an interview with the "Lord of the morning." The excess of weight had gone, the face was compact, smooth, lively in color, alive with an energy of soul the sparkling eyes no less revealed. There had come a keenness of appetite, and such sleep as her earliest childhood only knew. For more than two years there had been no hint that a headache was possible, and the injured spine no longer complained, by way of the head.

Ah, gentlemen and ladies of the medical profession, you are not getting such cures of aching heads through specifics. You are not getting such richness of blood through dosing with iron. You are not adding to your enfeebled stomachs such power of digestion with your array of tonics, coarse or refined. You are not getting such hunger for good food by sending your patients to the seaside, accompanied with only lazy ideas as to the stomach, as an organ of digestion, or the stomach as a lunch-pail.

Now, my laden friends, I am about done with you. In conclusion I can only reiterate somewhat as to eating habits. You cannot possibly err as to eating too little, or in getting those meals too widely separated. You cannot fail to get *immediate results for good,* if the means are duly applied. You will find the self-denial less a tax than you imagine. You will find, as you go to those meals with keener relish, not only added pleasure above what you have known before, but you will also find by studied effort to eat thoroughly ; and a determined effort to stop eating before full satiety ; that the smaller meal will carry you over more time of freedom from all sense of physical want than the larger ones you have always eaten. And I reiterate that Nature can make out her own bill of fare if you permit, hence you can have no need whatever to study the science of food analysis.

No food is wholesome, for all alike ; no food is ever to be eaten because, as a matter of theory, it ought to be wholesome. The old saying, " that one man's food is another man's poison," is based upon a physiological fact and one of most exceeding importance. My three sons approach the dinner table every day of their lives after a forenoon of rest for their stomachs ; the youngest son is very particular that his meat shall be of the fatty kind, even in his earliest childhood he would make his preferences known in this. His older brother is just as particular that there shall be nothing of the kind in his, while with the third it is a matter of indifference.

Do I need to spend any time at my table as to the relative amounts of albumen, and the hydro-carbons in the foods to be eaten ? With these hungry boys is nature likely to make any blunders as to the foods needed to balance the books ? I think not.

The darkness of ignorance you will dissipate, then, through the light of science.

I have tried to give you the key to the situation. Go on then and " work out your own salvation," not " in fear and trembling," but in hope that thy days may be longer, brighter, purer and happier " in the land the Lord thy God giveth thee."

LECTURE XVIII.

EVOLUTION IN CURE.

HEADACHES AND ASTHMA CONSIDERED.

My Friends the Readers :—

I expected to make Chronic Alcoholism the subject of my lecture this morning, but on receiving the following note I changed my plan.

Dear Doctor :—

I think all your readers were interested in the elaboration of your theory of the origin and evolution of apoplexy, and especially of your theory of its prevention, through the higher method in life-habits, and its possible cure by an evolution in reverse. But, deeply as we were interested in a subject so important, there are many of us who have more reason to be interested in the cure of those headaches that had been a torturing experience for forty-seven years. Indeed that an ailment that had become so imbedded as this must have become, and all the more intractable seemingly by a reinforcement from an injured spine, it does at least seem incredible that, after so many years and the patient in her sixtieth year, there should have been a miracle wrought, for it seems no less, and without even the " laying on of hands."

Many of us know all about headaches by personal experience ; we also know that there is very little to be expected from any treatments we have ever tried except temporary relief—*relief and not cure.* We therefore will be more than glad if you will give us any additional light on the origin, evolution and cure of this very common and often excessively painful malady.

Signed,

MANY READERS.

I shall most cheerfully comply with this reasonable request. In my account of the case mentioned I failed to let you know that there were only two subsequent severe attacks of headache, after the evolution in reverse began. These were followed by a succession of lighter attacks until the arch enemy was reduced to the condition of giant " Pagan," he could only growl and make ugly faces at the passing Pilgrims.

The attacks merged into mere growling hints that, in other days, were the mutterings, not of the distant storm, but of the immediate impending hurricane that agony only describes. The hintings, always scarey at first, finally " fainted " slowly away, and with the relieved head was also newness of life generally, because there was also a regenerated body.

Headaches, when not due to injury, are the evolution of an hereditary weakness, and are subject to the same evolutionary laws of development as characterize all other ailments. There will then be no need for me to go over again the science involved in the cure.

Children are fond of the illustration of pictures, so you get a deeper impression of the essence of a theory if it can be illustrated with a pertinent case. Having arrived at the very definite conclusion that all headaches, with the exception mentioned, are the evolutionary developments of avoidable mal-nutrition, I began to cry out with more and more emphasis against the sin of avoidable cause.

Now I had one martyr on my circuit, the mother of a goodly family who not only had these roof-raising attacks with increasing frequency, but was otherwise in a low condition. In a mental way she was a "last rose of summer," and her harp unstrung swayed to the waving of the willows only. Her cheeks covered with a skin void of all hint of life, hung down with feeble grasp ; the dull expression fitly represented the oppressed soul. There was all the look of "Leah the forsaken" in every one of the dull lines whereof the soul is revealed.

There seemed to be nothing in husband, sons, daughters, friends ; nothing in sunshine, bright days, brighter flowers ; nothing in all the air ; nothing in all the earth ; nothing in spring with its inspiring resurrections from the dead ; nothing in summer with all nature a Paradise regained ; nothing in autumn tints ; nothing anywhere in anything that made life other than an existence to be endured only.

I had been this lady's physician for very many years, and those headaches went right on ; there were those three daily meals, small in quantity, duly, habitually taken except during the attacks, and always with a sense of need in which hunger, relish had very little part. And I was simply powerless for years to arrest, in the very slightest degree, the evolutionary process.

But the time came when I could begin to talk. But talk for a long while was a fruitless spending of my time and eloquence. There was too much pressure from without : " this going without meals is little less than a crankism that borders on lunacy." There was

15

environment without, environment within, as strong head winds to beat against. "You must eat whether or no, or you will get so weak that you can't eat." Thus said the voices of interested, sympathetic, heathenish ignorance, with their counsel that tended not to "life unto life" but to "death unto death," and that over the jagged thorny paths of the martyr's way.

Said a physician to one of my patients whose feet had become established in a sure foundation, "I would advise you to have a *care* that you do not carry this new method too far, lest you become dangerously weak and reduced." Said this patient to me years later, "I wish I had this physician in my care that I might save him from the wreckage that is going on"—and that ultimately ended in death—perhaps premature by many years!

A time came in my own evolutionary history when I could address my martyr with all the force of intense conviction, thus: "You have simply got to enter upon every day of your mountain of tasks with a stomach void of even a grain of food, and when the time comes that you feel you must eat, that there is some food that you imagine will relish, then, without one glance at the clock, you must stop all and get it. There is no other way under the heavens whereby you can have these headaches cured, and you yourself get back up into this world from that uncovered grave in which you have so long resided."

My speech did not fall upon stony ground this time, and it was not so very long before a 10 : 30 A.M. breakfast and a light evening meal with the family could be taken with the keenest of relish, and of such strong nourishing food as she had not been able to take for years. And there began that same external, internal evolution, physical, moral and mental, that we see in

the material world when dead nature is released from
the icy grasp of Winter by the sunny airs of Spring.
The attacks became wholly relieved except when there
would be a temporary return by the special invitation
of the temporarily abandoned method. And at such
times there was no rush after me for a reiterated
lecture.

There is always a congestion of the blood-vessels
incited by pain, and the head is a most unbearable
place for such an abnormal condition whether within
or without the cavity of the cranium ; and the head is
the part of all parts of the body which should be
guarded with closest care against such conditions.

One lady who used to be afflicted with a severe
attack nearly every week has been able to almost
wholly relieve them, and she found, after two or three
years of study of her digestive powers, that one substantial meal daily gave the largest freedom from
head and other ailings. When one once begins the
study of digestive conditions with an ailment as the
inducement, *self-culture* is certain to go rapidly on.

Experience becomes a severe educator but an effective one, and it is no small matter in human existence
that painful experiences are so quickly understood as
the immediate avoidable results of violation of digestive conditions and not the symptoms of an unseen,
unknowable foe that can only be cast out by the ex-
orcism of remedies. In proportion as we have good
health so do we enjoy this world as an abiding place,
and he or she is in a state of disease who is in a hurry
to get out of it before every faculty, every taste has
become torpid through age.

Your headaches, my readers, are a product of your
own culture, and you will never get relieved of them
until you induce a culture in reverse. Those blood-

vessels that supply the aching structure in all headaches by frequent distensions have become permanently
enlarged, and the bromides, the chlorals, and the
newer specific remedies will never reach them with
toning power. *They will benumb the aches but enlarge the cause.*

One of the most inconvenient, annoying, and often
very distressing locations to have chronic distension
of blood-vessels is in the small bronchial tubes, to
which catarrhal asthma is due.

The condition in this disease furnishes us a fine
opportunity for a physical demonstration of the part
which dilated blood-vessels play in the development of
distressing symptoms that can be almost seen with
the naked eye. The blood-vessels of these smallest
tubes up to their termination in the air cells are weak
at birth through heredity. They began to slowly enlarge, perhaps only in a microscopic degree early in
life, as you have been told is the case in other locations,
through mal-nutrition from irregular feeding.

Little by little they develop, and a time is reached
when there is a slight catarrh developed and its associated cough. The water of the thin blood gets out of
the dilating pouches with more ease because of the
slowness of the circulation. At night the discharge,
the secretion as it is called, oozes out very slowly and
becomes thickened and sticky by the evaporation of
some of its water, and hence often is extremely difficult to relieve by reason of its adherence to the tubes.

In due course of development the minute blood-
vessels of these tubes become subject to attacks of
congestion that causes a contraction of the caliber of
the tubes, and hence a shutting off of the cells and a
siege of asthma, or of asthmatic breathing. As these
attacks become more and more frequent so do they aid

in a breaking down of the general health. Induced by avoidable mal-nutrition they become most potent factors in its rapid increase.

And it is a disease of so distressing a character as always to invite the most drastic treatments and remedies, with every dose aimed at a symptom only. Thus origin and evolution of the disease.

Now I am able to give you an evolution in reverse of a very striking character. It so happened that when my friend, the publisher, became inducted into this science of life he began to immediately think of his suffering friends, and one of the first thought of was Dr. J. II. Alexander, a dentist of Mystic, Connecticut, who was a great sufferer from asthma or asthmatic consumption, as some of the physicians believed it to be. There was a severe cough and the associated general debility that had made him unable to even dress himself without assistance during four years. He had endured all the misdirected dosage, special, regular, irregular and specific, that seemed to incite any hope, and for all his efforts he slowly got down to a hopeless and most distressing condition with death, apparently, close at hand.

My friend consulted me about this case and was assured that a rigid adoption of the method would at least prolong the life by affording some relief, and beyond this nothing was promised. Fully posted in the theory and an enthusiastic believer, he visited the home of this friend and spent a night with him. He was directed to enjoin a fast, no matter how wasted the body, until there should come keen hunger; that this could be done with perfect safety because the brain would get its due support from the tissues of his body that were of little account to him anyway, as he was totally disabled from labor. He was directed to inform his friend that, instead of a reduction of strength, he might even expect, not only an increase of physical, but mental strength as well.

This man, an ex-soldier of the late war, had got reduced to 97 pounds, and so the skeleton was put on a fast! And a fast that was to go to the eighth day! Did he starve and send my friend into confusion for promulgating a heresy? The result—every day a cleaner tongue, a declining cough—lessening of asthmatic trouble, increase of mental and physical energy. On the eighth day there was a call for toast, on the ninth, after a short ride, nature would have nothing less than a hearty dinner! And ever since this began, June 6, 1894, there has been a gain of general and local strength, and an addition of 12 pounds of muscle, and ability to go upon light duties.

That there should be an actual gain in those dilated vessels when nothing was going into the stomach but cold water seems very anomalous, but the blood was kept duly enriched for brain purposes, by the tissue supply that was not nearly exhausted; and there was no diversion of vital force to aid in the chemistry of decomposition. And when the stomach had rested, had grown into power, as the grass grows when favored with due atmospheric conditions, it did not send up the signal in a faint, indistinct, still, small voice. Under the favoring conditions of increased nutrition those little vessels began to get toned up, and the secretion diminished, and hence it was not possible for those congestions to occur that prevent the air from reaching the cells. And so there was an evolution in reverse, no less general than local.

One more case. A lady came to her bed after an exhausting journey, with a light fever. She too had been a martyr to asthma, and with a husband with

great executive force and marvelously hopeful of un-
tried treatments never abated by failure ; 1 think that,
through his efforts, no case of asthma was ever more
drastically treated, and for nearly a score of years.

In this case the sense of smell had been abolished for
a long time, and the sense of taste nearly so.   These,
with a catarrhal condition of the throat were the only
marked local complications.

My patient was at once put on to a strict fast with
not so much of an interruption as a swallow of any of
the meat teas.   And this was continued for fourteen
days.   On the fifteenth day there was a degree of
strength achieved that permitted the use of the arm-
chair for a few hours, during which she took some
broth with keen relish, as if there had been a new
gustatory nerve inserted during the vacation.   Now for
a very surprising statement.   A few mornings later,
and after some substantial meals had been disposed
of through the chemistry of digestion, as she was
quietly enjoying her arm-chair in her upstairs room,
suddenly there was a violent attack upon a resurrected
olfactory nerve, and with an odor that to her was fearful, and the first hint she had had for years that there
was any such nerve !   And what was the matter ? a
door was opened and the odor was found to be from
boiling coffee away down and back in the kitchen !
The improvement was general and rapid, and for more
than a year there were no hints of asthma.   There
came trouble later, but it was due to the avoidable
evolution of development.

Ah, gentlemen and ladies of the medical profession,
you easily admit that no such results were ever reached
by medical treatments, whether general or specific
in this disease.   There is at this hour a reverend
gentleman in his New England home, with an over-

stock of predigested food, who has his wrestlings every morning to dislodge the sticky, evaporated mucus which, with tenacious grasp, clings to those little tubes, and narrows the way whereby the fresh air of heaven bears its unseen oxygen in a labored way to that foe that must be instantly paralyzed, or agony beyond conception, and death is the result.

He has begun to deal with all his acquired ills by an evolution in reverse. Little by little his weight will go down, and so will those little tubes and their life-channels get toned into the normal size, and there will be no more need of relieving inhalations. And there is no other way known to man whereby he can be saved. There is no other disease, local in its character, which so laughs at, which so mocks at all human efforts to cure with remedial means. But Nature will have her time and all the time she requires, when she is permitted to conduct an evolution in reverse; she will be exacting as to every aid, will resent instantly every hindrance, and she will deal with those local diseases on the divine plan with none of your disturbing assistance. She will place the new cells just where they are needed, remove the old as fast as they have outlived their usefulness, and the "waste places" will be regenerated unless the sin has been beyond redemption.

The cure of a local disease is the growth of a new part in the same sense as the growth of the whole body from the single cell. That is Nature's own work, and in no sense a matter of specific treatments.

I will close the lecture by giving you a physiological fact about the brain that should have been mentioned further back, but which will not lose any of its significance by standing alone.

The brain is capable of sustaining its functional power in a very high degree on very thin blood.

I once saw a patient in the last degree of waste whose mind was as clear, it seemed to me, as in the prime of life, and yet the blood was so very thin that its water was constantly escaping into the cellular tissues of the feet and limbs, which were badly enlarged thereby. This may be accounted for by the fact that only a small amount of nourishment is needed by the brain as compared to the whole body, in a state of activity.

LECTURE XIX.

My Friends the Readers :—

I have invited a few of you younger victims of the drink disease to hear, not a lecture, but a talk on its origin and development, and a natural method for its cure. I call it a talk because you have been lectured until you cannot hear the word uttered without experiencing a grating sense upon the "spirit's finer ear." Whatever eloquence can do to lacerate conscience in its portrayal of wrecked homes, with their children shivering in rags, and crying for that bread of life that you have handed to your able-bodied fellow-citizens, the well-dressed pictures of health, across bar-counters, has been done over and over and over again, and still you drink on all the same. Mathematics and statistics have been arrayed against you all in vain.

There is nothing of argument for you in homicides, in incarcerations, in jails and prisons, because of reason dethroned by strong drink. Moral energy spends its force to little purpose when directed at symptoms and not at causes. It is of little avail to lecture the proprietor of a jumping tooth on the impropriety of being awake at the midnight hour as the destroyer of the sleep of his entire household. He fully understands the virtue of that sleep which the " just " are supposed to realize in the highest degree ; he fully realizes what every groan is to the exhausted wife, or perhaps to

234

the invalid daughter, but he will groan all the same. His pressing need is not a lecture, but the forceps or a benumber to the inflamed nerve.

Eloquence on the platform, in the pulpit, in the journals is all leveled at symptoms, and in vain. What you want is the helping hand. Yours is a disease that has gone through its history of evolution step by step as have your catarrhs or your asthmas. You cannot stop your aching tooth by will-power, nor can you deny the consolation of a drink when all within is on fire. You drink, not because of a perverse moral condition that you ought to control, but for the anæsthetic ease that raises you, for the time, from the depths of misery to perfect comfort and most enjoyable indifference to all corroding ailings or cares. Yours is not just one tooth but a universal aching that will return whenever the dose has lost its soothing power. Agony demands relief, not argument.

I have invited you especially, because you have enough moral power left to have a strong desire to be cured, and also because you are so mentally and morally endowed that your lives are worth saving.

There are very many drinkers who are scarcely worth the effort, because they are the floaters of society, without any tangible excuse for being alive. But you, you with your larger powers and those young children growing up about you, your lives are worth saving, for what will be the result to yourselves, to your families and to all your fellow-citizens with whom you are brought into relation.

I have invited you to bring your wives along that they may get some ideas as to the development of this disease which may be a help, a warning, perhaps, that may avail in a better training of those children.

My unfortunate friends, you will get no words from

me that will have the slightest hint of reproach that
you should have permitted yourselves to become so
diseased.

The drink evil is not confined to classes, nor to minds
naturally perverse in morals.   There is a frightful per
cent., even of clergymen, who are also chronically sick
even as you are.   And perhaps the heaviest per cent.,
considering culture and knowledge as a basis of
classification, is to be found in the medical profession.
Physicians are tempted beyond all others to seek the
aid of temporary relief from the pangs of exhaustion.
I tell you, when you keep your physician hour after
hour when Nature designs he should be asleep, because
he has already had his day of taxing labor, when you
keep him hour after hour through the long nights, a
victim to his own intense anxiety over your life or
death case of sickness, and all the more a victim
because he is the center against which your harassing
anxieties are constantly beating, if the drink disease
has any hold on him, he is going to take a drink for
the same reason that you do, and his larger knowledge
of the evil of it will not deter him in the least nor for
one moment.

With the majority of the medical profession pre-
scribing alcoholics from their supposed powers to sustain
vital force in attacks of disease ; and with such a heavy
per cent. using them for the precise reason that you
use them, and with as little will-power to resist the dosage when the need becomes pressing, I cannot, with any
consistency, spend time over suggestions that you
deserve all the suffering that has been yours for being
so foolish as to permit yourselves to fall into the
relentless hands of this chiefest wrecker of souls as
well as lives.

The foundations of the alcoholic disease are begun

in the nursery, and very generally during the first few hours of life. It is my conception that there is a distinctive alcoholic temperament born with many of us that reveals itself in a nervous system that is peculiarly sensitive to disagreeable impressions, and a relative lack of that stoical indifference to ailings, whether mental or physical, whereof the Indian is supposed to be unusually endowed.

I had once a family in my circuit where the husband would be at his daily employment without a break, but when the wife would get sick he would get on a drunk, and prolong it as long as there was danger in the house. Why such a giving way when he should, of all times, and for her sake, be in the most sober condition ? Let me tell you that, away down in the depths of that laborer's nature, there was a refined sensibility, a sense of what it was to him to have the wife dangerously sick, that became *pain unbearable*, hence the anæsthetics, the benumbing poisons. And he yielded just as others would do, if a jumping tooth were the coincidence of a sick-bed, and the dose would be taken even if it were to paralyze the acute sense of the needs of the sick.

My friends, there is not so very much difference in what any of us will or will not do, to relieve agony.

Now, young mothers, I must address myself to you. It is my impression that there is not a mother in all America that has a clear conception of the easily-ascertained conditions involved in the digestion of a meal. There is no idea of the need as it actually exists, that, after a meal has been digested and the indigestible residue has been cleared away, the stomach must rest for a time undisturbed, until the little glands become able to generate and throw out their dissolving juices for the next meal. There is no idea that the amount that can be drawn out at each meal deter-

mines the amount that should be offered for solution ;
no idea that the excess of this supply to be dissolved
is the prime factor in human wreckage. It is not
known that when feedings are widely separated and
given at exact times, that Nature herself makes the law
of supply and demand with almost unerring certainty.

There is not even the remotest conception in any
mother's mind that the brain of an infant will be in
a better condition in the morning, if the stomach has
rested all night ; not the slightest conception that the
wreckages of the night are all due, not to the lack of
brain nourishment, but to brain-taxing through the
chemistry of food-decomposition.

And so, within the first twenty-four hours of life,
begins the life-struggle against the power of death
through the darkness of the nursery.

Irregular feeding incites still greater irregular feeding, and then come the soothing syrups, and, as a result,
those of your children with the peculiar nervous alcoholic temperaments never know what real comfortable
existence is ; they require a great deal of care, their
wants are in perpetual motion, and very largely their
lives have become an afflictive dispensation. Not
happy themselves, they see no reason why others
should be happy.

Being among those of the " fittest," they get through
the first year of a miserable existence with their lives,
though grievously sinned against. Getting to an age
when they could do their own feeding, the evil work
goes on. With apples, nuts, candies and between-meal
piecings, those stomachs are never allowed to be
empty. They never sit down to their regular meals
with any satisfaction ; they never get up from them
in that supremest mental and physical condition when
all sense of need is abolished.

Young manhood is reached with the nervous system ripe for that first dose from a bar tumbler, and all at once there is a soothing, a "peace-be-still" influence upon all within. The match has been struck, a small fire has been lighted, and the end is, physical, mental, moral death !

What has been going on in the mucous lining of the stomach all these years ? My other readers have been told that an alcoholic is an irritant to this membrane, that it causes a congestion of the little arteries, capillaries and veins, thus subjecting all tissue between them to life-depriving pressure, which also involves the gastric glands by which their functional power would be lessened.

This goes on month after month, year after year, until this membrane has become an alcoholic wound ! But food is sent down to it, a little is digested, and then comes the tax of decomposition, and with it the need for the soothing drink, and it will prevail. With stomachs in such a condition, so diseased, can the voice of reason prevail, no matter how eloquent, how persuasive ?

Said a man to me who had had a large and long experience in the use of alcoholics in their habitual, daily use, " I had to have a drink to give me an appetite before sitting down to my meals, and then a drink immediately after each meal to allay the nervous irritability incited by the digestive process ;" and why should not this have been with the starved, strangled glands, with their walls so dry as to fairly crackle under the movement of enfeebled peristaltic force ?

Ladies, in that need of stimulation that comes from exhausted vital power, exhausted from nights of taxing labor, aggravating an hereditary digestive weakness, acting upon the peculiar temperament by as much easily roused to an overpowering appeal for instant relief, my need for the elevating, soothing cheer of the cup of coffee has often been resistless in its force, and all the more these years when I have the pressing need to avoid all professional work at night. I know full well what it is to raise this cup to the lips against the moral sense of its sinful use. I too have had all the illustration of the difficulty of getting back from a long straying from the paths of hygienic virtue, that I need to fully understand what it is to want the habitual dram.

And I know also what it is to enjoy perfect freedom from the sense of all need of the cup that cheers, that inebriates, but does not drunken. And for me there is no safety but habitual, total abstinence if I am to be free from an intolerable condition of the nervous system that will appeal with resistless power when the need is incited.

That I have not been in a drunkard's grave for years is wholly due to a father's illustrious example, who, in his young manhood, had the courage to organize, with the aid of a few sympathetic souls, the first total abstinence society in his state. Courage? Yes, it always requires courage to be in advance of the people in any reform. In those earlier times when even the clergy might indulge the social drink in moderate degree, it required courage to take every step in the direction of total abstinence. He had need to keep his whole stock of courage ready for the offensive attacks of ridicule that were his for many years, as his son had need to care for his in the reform that sent him in advance of the people. Had the match been lighted in my case to fire the alcoholic train, hereditary and acquired weaknesses would have paralyzed all resistance, and death, through the martyr's thorny way, would have been the sequel, premature by many years.

My father's father always kept his bottle of liquor, and was as temperate in its use as any man well could have been who used at all. And why did he keep it at all? The universal custom of the time made it a necessity. Necessary? Why?

My father often used to relate that there was an unexpected call from several clergymen one day, and that, by an over-sight, the whisky bottle had been allowed to remain empty. He was the only available son, and was seized upon to go to the nearest bar and with speed return, that due courtesy should not be unduly delayed, but declined. To my father's illustrious example am I indebted for all the life I have had that has been worth living.

Young mothers, you can have no higher mission on earth than to so care for the eating habits of those children as to keep them above all between-meal hankerings and wants, that they shall be safe from the temptations of the spoiler when they are out from beneath the wings of your sheltering care.

"Lead us not into temptation." Is there not a suggestive hint in that, that if you get into temptation you are going to yield? Young mothers, you have no idea of how much you are doing to create those indescribable hankerings that shall make those children easy victims of the spoiler's hand. It is you who are, not leading, but driving them through the saloon door by a duly prepared culturing of the conditions.

There needs to be organized a new W. C. T. U. association, cultured in the science of a higher living, to engage in a new crusade, not against the saloon primarily, but against the all-abounding intemperance of the nursery, of the dining-room, against the intemperate excesses of the school-room, whereof the saloon is made possible, and whereof it is to remain a legitimate

16

result, no matter how the war against it is conducted, no matter how the attempt to restrict or abolish by legislative enactment. The clearly avoidable intemperance of the nursery, and of the dining-room, is responsible for more human misery, more of lunacy, of homicide, of suicide, of disease, of pestilence, of premature ending of lives, infinitely more than can be charged to the saloon, because infinitely more universal.

There must be a higher order of temperance in the home ; there must be sent into manhood, into womanhood, a new generation of young men and women, that there shall be no demand for a new generation of dram-sellers. Let the old organization go on in its fruitless war on symptoms and their inevitable, their logical results, the saloon of the present generation.

Let the new organization turn its attention to the regions of cause, the nursery, the dining-room, that there may be that culture of health which shall raise above all temptations ; and the next generation will have little need to war upon the saloon. Remember the aching tooth will have its relief. There must be dentists and forceps—there will be saloons unless legislated out of existence by home enactments.

Ladies, let me tell you a great physiological fact. Those children of yours can soon get trained into a condition that will enable them to go through an entire forenoon with more power of muscle, with keener intellectual action, with greater cheer of soul, with stomachs perfectly empty, than whenever a very small breakfast is taken.

Let me tell you another fact, that this habit once formed and continued into young manhood, there will be such a high condition of physical comfort between meals, there will be such habitual loftiness of cheer that a dram will actually prove such a depressing agency, it will be so immediately followed by a sense of discomfort that its repetition will not be invited ! !

The new crusade must be a war on irregular feedings. There must be incited a clear and an easily-ascertainable apprehension of the conditions of normal digestion.

The fact that the highest possible reach of physical, mental and moral perfection can only be attained by a total abandonment of between-meal nibblings and lunchings, no matter whether the innocent, enticing apple, peach or pear, by a total abandonment of tobacco, tea and coffee, must be clearly apprehended.

Will it be worth your while to start such a school of health-culture in your little families ? Organize such a crusade against the saloon of the next generation ?

Let me startle you with a statement. It is my belief that were every man, woman and child of my native city, with its ten thousand inhabitants, to be placed upon this physiological plan in life and carry it out with due rigidity during one year ; that even I, after my thirty years' wear in their days and nights over the sick, could easily care for all the cases of acute sickness that would occur during the second year, and without becoming disabled.

I have had a great deal of reason to make this statement from the cutting down of ailings in my families which have adopted this method in life-culture.

Will it be worth your while to have this blessed assurance in your daily life, that, as this higher living protects against the drink evil, so does it against all other physical evils ? That, as you have the happiness that comes from the preparation of that food that is going into empty, rested, vigorous stomachs, you will also have the greater happiness of seeing the work of your painstaking hands most keenly appreciated !

And, still more, there can only be the very least possible danger of disease, no matter what its form or character, when there is such obedience to the divine laws of life.

It has been very often suggested to me that, by waiting until this keen hunger comes, there is always the danger of over-eating. That is very true, but the rested, the vigorous stomach can very much better manage an overtask than a feeble one can, and then the painful experiences due to gluttony are sharp reminders of sin.

A great deal could be said as to the need also that the normal amount of sleep shall be duly cared for; of the necessity of guarding undue mental and physical labor from directly impairing digestive power. But the school of experience is a constant source of higher culture.

Then there is the problem of church suppers and other untimely, sin-enticing repasts that have to be most uncomfortably recovered from. In this let me illustrate. A few months ago one of my young men met a lively party of his associates at an evening gathering where, at the usual late hour, was served the ice-creams and cake. It was a young set that indulged, and with all the energy and impetuosity of youth. But there were sick stomachs very soon with several, and lowered mental vivacity with all except my young man. He was so perfectly well, even though his only meal had been at noon; he felt in such cheer, in such perfect exhilaration of comfort, that there was not enticement in foods so gilt-edged for him, and so none was taken, and his was the liveliest intellect of all on the " home-stretch." That evening's repast was the fourth that day for all the others.

If these life-depressing, sin-enticing church-suppers

are a necessity to the spread and maintenance of the gospel of the New Testament, then in the name of the physiological gospel, of the science of life, let them all be the second meal after, long after, the first light one has been taken, that their evil may be reduced to a minimum.

I will now let you all go to your homes to do some most serious thinking, and when you meet me again, I will see if we can make any attack on the present generation of dram-dealers that they will become conscious of in your individual cases. Whether there is any helping hand that is attached to an arm long enough to reach yours as you lie helpless, prostrate in the ditch of our human highways, too very sick to be raised by reason, by denunciation, by eloquence, no matter how tender, how persuasive.

LECTURE XX.

CHRONIC ALCOHOLISM—(*Continued*).

THE TREATMENT OUTLINED.

My Enslaved Friends :—

My experience in the cure of such diseases as yours is rather limited. For very many years after adopting my present scheme for the aiding of nature in the restoration of her waste places, I had no thought that there was relief for your afflictions. There is the one difficulty in the way of all treatments, a difficulty that confronts me now in the treatment of your cases. And it is just this : you have the constitutional right to commit slow suicide and I have no legal right to restrain you. Could I have legal power, I should have no difficulty in getting you as completely freed from the toils of your enemy as you were before you took that first drink ; and once released from his grasp you would be able, through a higher culture in obedience to life's forces, to go on in the ways of your affairs infinitely safer than you were when you left those homes with the parental blessing.

You, my friends, are sick men, very sick ; you are prostrate, helpless ; and you have no rest from that foe that gnaws like a file and stings like an adder ; not one moment of blissful oblivion from these filings, from these stingings, that does not come from those doses that make oblivion of all care of your vital concerns in life.

Will you drink when you are in this mental and physical agony? Yes, even though you snatch the bread of life from the hands of your starving child that it may become the drink of death for you! Moral suasion, argument, denouncement, ridicule, dashes in vain against such a rock-faced fact as this. You have all tried the " gold cure," and you thought for a time you were cured ; but you were taught no ideas while you were patiently permitting your bodies to become needle cushions in an absurd treatment, no ideas as to the science of prevention.

Your residence at the hospital was a relieving break in the hard lines of your life, and your nutritive powers for the time were stimulated by the inspiring power of hope. You all expected to be cured, and there was a powerful appeal to endurance from the forces of association. There was a great deal of relief achieved that was due, not to the needle, but to a combination of forces to which you were subjected. And what is the difficulty that now stands in the way of your getting well? Let us look into your stomachs for a moment with the " mind's eye " to give you another and fresher view.

The stomach is in a state of chronic congestion, and its glands, as I told you in my last lecture, are dry, or so nearly dry that they are never able to meet the demands for the work you habitually, irregularly and excessively place upon them. What are you going to do in such a case?

Now, for instance, suppose that those stomachs, in perfect condition, should be subjected to a violent rasping from a grater, which would convert the entire mucous lining into a wound. What kind of treatment would common sense suggest in such a case? Why, of course, the " rest cure," as must be in every other wound and in all fractures. Your stomachs, every one of them, as I have suggested, are like the school-yard in term time, there is no growth there, nor can there be until the vacation comes. Can you give your stomach a vacation? It will never recover without it, nor will you ever be relieved from all your resulting ills until they do recover.

You can easily see that a dosing is not likely to relieve a condition that is being kept up by irritation; and you can also clearly see that the changes that must take place in those stomachs must be through the divine hand of nature, even as in the case of wounds and fractures. Now, how are we to get that rest that must be if a cure is to be wrought?

Let me make a startling statement. It is my belief that the alcoholic disease is one of the easiest in the world to be rapidly and permanently cured where the life is actually worth saving. That is, in all cases where there are powers which are worth the effort.

There are a great many adults in this world who have never become men and women in any manly or womanly sense. They are children, and the most child-like in every moral and mental sense, and they could all be spared without causing any vacuums in society, and with the result of clearing the social atmosphere. How so easily and rapidly cured?

Now suppose you have made up your minds to give those stomachs a vacation, which will be a great thing as a starter, you will then expect the following evolution to take place: as soon as the stomach becomes entirely freed from every grain of irritating food, it is going to begin to get well; as soon as the bowels have become entirely freed from decomposing food and indigestible residue, they are going to rest into power; the entire digestive tract in a condition of repose, there

will be no more exhaustive demands upon brain force.
This condition once reached, the cure goes on rapidly.
Now comes in our wonderful bill of fare to help you
out. While the fast is going on the brain will duly
care for itself, and far better than it has been able to do
for years.

Now let me startle you again. The digestive tract
once empty, every moment of the fast will be a moment
of unloosing of your enemy's grasp, every moment
will add to the relieving sense ; and they will add a
momentum to the cure—there will be an awakening,
a growth in all the forces of life.

And as the fast goes on by as much, will it easily go
on, and at last when that hunger comes such as you
have not known for years there will a rested, perhaps
a cured, stomach ; and there will be such an utter
absence of your old adder and file agony, that you could
not even be hired to take a drink of whisky. You will
have no want other than for good, nourishing, substantial food ; and for once in your lives you will have
clean tongues ; and that nerve of taste, new-born, close
to the surface, will give you a new idea of what it is
to eat with that relish, that is always associated with
the power of selection and power of digestion.

Can you fast, do you think, until your stomach is
rested into life again? If you cannot, because will-
power has become too weak for your human effort,
then I cannot see you cured until I have the legal right
to seize upon you and arrest the progress of suicide.
There is no other way given unto man whereby that
stomach can regain its power, except through the
vacation that must be voluntarily or legally enforced.

The long-suffering, long-abused stomach may require a few days, or one, two, or three score of days to
reach its normal power ; but it will be one of the

safest of all human methods to relieve it, to cure it,
because of the absolute safety of the brain while it is
going on.   You will only lose the relative proportion
of the tissues that can be easily spared ; and lose
without the slightest sense of their loss except what
appeals to the sense of vision.   You will see that the
loss is going on, but cannot feel that it is ; and mark
you, while this process is going on, the saloon is losing
its grasp on those revenues that deprive *your children
of bread!*   If there be any other effective way to
attack the saloon of the present generation, I do not
know what that way is.   If there be any other way to
cure a diseased, pickled stomach, other than through
a perfect fast, I do not know what that way is, any
more than I know of any other way to cure a fracture
or a wound.

That it is a safe way beyond any question, I do not
believe any sane man or woman will care to deny.
That it is a physiological method in its very essence I
do not believe any sane man or woman, whether of the
profession or of the laity, will care to deny.

And what about the second meal after the first that
you have relished so well ?   You are going to have it
after another fast until hunger comes, and so long as
you shall keep this up you will not have even the
slightest disposition to make glad the souls of the
present generation of saloon proprietors by your
patronage.

In due course of time you will find, as in all other
diseases, the development is general, and that all other
weak or diseased structures, alike improve with the
stomach.   And you will also find that, if you take due
care of the stomach, the bowels will care for themselves without any artificial aids.   Nature has her own
disinfectants, and she has no need of your goads ; no
harm will come if they are not used.

On one bright Sunday morning in June, 1894, I delivered, by special invitation of a wife, an address on temperance to a husband who was sick. It was to be on Sunday morning, because I would find him at home, and also because he would not be sober enough to listen to an address any sooner. The patient I found to be a young man of splendid physique, and at the very entrance upon his ripest maturity. He had a stomach vast in its powers of accommodation without complaint. Abuse unutterable could be heaped into it, but there was no outcrying from anguish. And there was a wife and a fine group of children.

During every morning of their married life, when not disabled by sickness, the wife had prepared an elaborate breakfast for her family. The man was a large beef-eater, and never could he go out and draw in the fresh air until his stomach had become packed. This would not hinder the want of the mid-forenoon ham sandwich with beer, nor a mid-afternoon deal of the same kind after the hearty dinner.

And so life had gone on with the taxing of vital power that would come from several irregular meals daily and untold potions of whisky and beer. I met him on this Sunday morning not only a wreck himself, but with his business in a general wreck, because his judgment had been dethroned. It was, even for this reason, a desperate case to encounter.

I did not enter upon it with large faith, but I did get a fast enforced that was attained by the evolution I have outlined ; and there was actually no want of the relieving dose when that first meal was taken, and without tea or coffee.

With his powerful natural digestive and assimilative forces the relief was rapid. I saw him daily for some weeks, and I know that there were only the two moderate daily meals, and his assertions were strong that he was continuously without any desire for the saloon ; and every line of his countenance was an assurance of regeneration. I saw him several months later, and though he had been without steady employment, he had only slightly broken over two or three times, while the reform method in diet had been closely adhered to. There had been a marked decline in his weight. In this case evolution is likely to reach a perfect cure, at least a great deal of prevention, for the old eating habits will never be resumed.

And what about the family ? I got that same reform started that I have been talking to you about and with the happiest results. What is it not worth to the mother of a family to know that the wastage, the taxing, the time spent over an elaborate breakfast is wreckage to health ? A wreckage that makes disease and the saloon possible ?

In line with this rest-cure theory is a case that finely illustrates it. Three years ago I made a short address to a man who, for more than twenty years, had been a sufferer with a disease stomach. Not finding any aid in America, he crossed the ocean, and for eighteen months was under the care of a specialist at Brussels for chronic gastric catarrh.

When he fell into my hands it was his abiding habit to have his enforced breakfast ejected after remaining " in soak " for a couple of hours, when the little period of rest from thence until the noon meal would be followed by the ability to retain, in an indifferent way, a light dinner. But the evening meal seemed to always prepare the way for the usual morning " heave offering."

Now he did not consult me as to this case because, with failures by experts, he would scarcely expect that an obscure physician of a little inland city would be worth his while to spend any time over ; but the coincidence of my first professional services in his family brought his case to my knowledge.

Satisfying myself that his difficulty was more an overtaxing since his birth than any real disease, I suggested to him that the taking of a breakfast that could not be kept down, actually amounted to a very disagreeable fast during the forenoon, and perhaps a fast that should last all the forenoon without a spell of sickness might be endured, with as little loss of strength ; certainly, if no unbidden breakfast were sent down, none would be sent back, with a " slammed door" emphasis. He got keenly interested, at once seeing the logic of it. He started in the next, morning and got through all the forenoon with more strength and cheer, because he was not sick during the forenoon, and when that dinner-hour came there was not the least sign emitted, that he had struck the wrong time of day for his first meal. And from that day on ever since he has kept faith and without a single reminder of his old enemy "chronic gastric catarrh," and ever since he has not failed to wonder over a cure so simple, so marvelously rapid.

Ah, readers, it is not myself, but Nature I am holding up to you, and Nature in her own colors, there is no fiction, no " Beplastering with rouge her own natural red."

I was no less surprised than was the patient, and he is not over his surprise to this day. As soon as he found himself among the saved, he induced his wife to adopt the plan, and they fixed eleven A. M. as the best hour for the first meal, with a second one in the evening, and no meals have been missed by either in three years.

There chanced to come into my line of life a man who had had gastric catarrh for years without the ejectments. He had nasal catarrh also. He was a man of wiry build, of strong muscular fiber, and withal a man of unusual ability and acuteness of mind. He became at once keenly interested, and being a scholar, especially interested in all matters physiological—physiology having been a favorite study with him in younger years; he began to watch developments generally and locally, with an acuteness and intelligence I had never seen in any other person.

I had called his attention to the demoralizing effects of tobacco on the sense of taste, as well as on the nervous system. Now, for not less than twenty-five years, this gentleman had not known what a natural appetite was, and he used to often wonder and regret, that the man could not have the appetite of the boy. He was scarcely a patient of mine, but as we were daily thrown into each other's society by virtue of a business relation, his growth in the grace of health had to be subjected to the slower evolution of a gradual scaling down of the morning meal, not being inclined to a radical change ; but the scaling down went on and the health improved, and soon he began to complain that his hunger was becoming unmanageable, that he was constantly overtaxing. The old-time boy-hunger and relish had returned.

Evolution went on, and no artist ever watched the development of a picture with keener interest than he the developments going on in his body, generally, locally, and in his mind. In the earlier days he could not spend an hour with me in my office without needing the cigar or chew ; but the former I could not admit, as to draw the breath of life untainted by tobacco smoke I considered one of my inalienable, constitutional rights that T. Jefferson has written very clearly about, especially when I draw it in my own office.

As for "cuds," I have never believed that spittoon-cleaning is any part of my business or of those about my household, hence the spittoon was abolished years ago. Nor have I ever thought that the odors of decomposing, ensalivated tobacco-juice were anything less than a violation of my own "constitutional laws," flagrant, nauseating, health-defying.

The cigar omitted by virtue of my views made known, the other was very soon abandoned. In due time my friend informed me that he proposed to stop the use of both, but that he would let it go for the express purpose of finding out whether the habit would not abolish itself by virtue of a lessening need through a better tone of the entire body, and a developing sense of protest from the sense of taste. And this actually happened. He reached a time when he was able to abandon both, without any call upon the will-power, and more than this, it was only a few days before he found that the nasal mucous membrane had reached a stage of functional activity that utterly revolted against the odors of tobacco, and no less, breaths laden with beer or whisky !

My sick friends, how did this gentleman reach this new-born physical regeneration ? It was a very instructive case to me from the fact that he had so keen an insight into the physiological basis of the regeneration that was going on, and also because of the reverse evolution that was going on in the tobacco disease.

In my evolutionary history of this new method of living the question of the use of tobacco was scarcely raised or even thought of until recent years.

I had my hands full to get the breakfasts scaled down, very full to get them abandoned entirely in special cases, and this was enough to give me all the notoriety of "crankism" in the superlative degree. Had I attempted more reform at first, had I included tea, coffee and tobacco as foes to human life, I have no doubt that I should have failed in trying to do too much.

It was not so clear to me until recent times, that the use of these apparently harmless soothers are after all only lullabys to Nature's complainings, a muffling of those voices that arise from disordered nutrition that should be stilled by a removal of the cause. In my own case in the use of coffee I have hundreds of times yielded to the tempting cup and so silenced those voices for hours at a time, but they would always be heard again.

And so all the more I can assure you, my sick friends, that there is a luxury in living when the cup has lost its tempting power, a luxury in being able to say "No" with no painful exercise of the will-power.

And how did he reach so high a degree of nutrition that this excessive use of tobacco in the two forms of the habit was relieved, with no tax on will-power ? It was only by following up closely the method outlined.

And now what are you going to do about it ? Can you resolutely undertake such a fast as I have enjoined and make it a fight for life and for all there is in life for you that is worth living ?

Some of you also have nasal catarrh, have lost the sense of smell, and none of you have the normal sense of taste, especially you who smoke and chew. Not one of you but has some local ailing that is easily located, while there are frightful possibilities of what may be in some of the vital organs.

With my assurance that the hand of disease may be

stayed as soon as the entire digestive tract gets into repose ; that not only will the disorganized stomach begin to grow into health as the grass in the schoolyard during vacations, but in all other diseased structures as well, will it not become you to make a supreme effort, the supremest of all your life? I assure you that as this fast goes on, there will be less and less need of your will-power to sustain you, but it must continue until you would prefer to take that first meal in the dining-room of the axman's camp that I told my other listeners about.

And then the evolution that will go on in the faculties! What will it not be worth to your wives if you can regain something of that felicity of disposition whereof you had an abundance on that first day of your wedded lives! What clearness of vision then to see the slightest hint of a want, what acuteness of understanding to know it instantly, what rareness of moral powers to make devotion a luxury that has more of Heaven than of earth in its essence!

Ah, gentlemen, those wives have been dragged down, chained to those "bodies of death," not to husbands. They are not troubled with catarrh, but the sense of odor has been blunted. They have had that sense too many years subjected to a combination arising from decomposing food, from the cigars and cuds, and from the distillations of beer and whisky ; and patience, long-suffering endurance has been the daily need—all this you have realized with more force than language can utter.

You are to make the effort, and after you arise from that first meal with the axmen, you will be happier than you ever were before since those early wedded days. You will not be tempted again if you duly space those meals. The victory will be yours if your life is worth saving.

17

My list of successes is small because of the short time in which this method has been under tests. I have very little faith that any method will succeed with adult children. There must be brains to appeal to, and brains that make the possessor worth effort, for what his life is worth to himself and to those who will be the happier for all the life he gains.

I succeeded in getting one man on to a fast who for years had been in the habit of using beer in extraordinary amounts. He was a fine mechanic, and, being disabled by his excesses, was thrown out of employment, and from thence every dollar earned, in often menial employments, was converted into drink. He had duly graduated from two gold-cure institutions, but only stayed cured for a few days after each. With very untoward conditions for starting out he succeeded in getting to his feet, and from thence on there was perfect health for him for nearly two years, and then getting away from home he fell among thieves who robbed him of his higher life.

A few years ago I was called upon to care for a man suffering from his first attack of delirium tremens. He had been a periodical drinker, was able to go for months without a single drink, when suddenly his lion which had been going about very silently would pounce upon and seize him with a roar.

He had an ostrich stomach, one that never complained from abuse, but only so much digestion could take place, and the surplus was crowded out of the way with no groaning, but the brain had to become the reflex interpreter of violated law. There would come those times of sudden overpowering, and then several days of dissipation. With no definite idea that a cure might be achieved, I advised him to be content with the very smallest breakfast, knowing full well that reform could be pushed no further with him. The result is that there have been but one or two attacks since, and not long or noisy ones ; and he has assured me that during these years he finally grew into a condition of feeling no desire whatever for the soothing drink, a fact that never characterized his total abstinence periods, before he began the better system of living.

The method will succeed in proportion to the radical degree in which it is enforced and maintained. I have had one patient who has been on my hands now for several months, who has had these periodical attacks for many years. During this time he has had only two light ones. And as the evolution goes on they will grow lighter and lighter, until he will be in no more danger than he was before his first drink.

There is a building-up process in this method that is instant at its adoption, and as the effort is radical and persevering, so will the results be more marked. There is nothing in "periodicity" that need to discourage, for there were no sudden attacks of need before the first drink was taken. This is only one of the rarer forms of the alcoholic disease that will disappear by reverse evolution, as it was developed by evolution.

Now, gentlemen, you have no need of gold-cures, no need of a residence in an infirmary for any specific course of treatments, if you can organize your scattered forces into a compact body and hold them well in line and in "close columns." If you cannot do this, then surrender your legal rights for a few days and forcibly endure what you have not the self-constituted power to institute and carry, and victory will come.

In closing I warn you that your case is hopeless unless those diseased, wounded stomachs are given that perfect, enduring rest which must be given in your fractures or your amputations; and a rest which will be used by the brain for adding life to its functions, and that will ask no aid at your hands, that cannot be granted without the sense of loss in any debitating sense.

At our next meeting I will consider several questions that have been raised by the curious-minded among you.

LECTURE XXI.

SEVERAL QUESTIONS CONSIDERED.

CONTAGION—QUANTITY OF FOOD PER DAY—THE BATH-TUB—REDUC-
TION OF WEIGHT.

My Friends the Readers :—

The several questions you have raised I agree to consider only, to answer is quite another matter.

DEAR DOCTOR :—

We are all agreed that it is very much better for us to await hunger before eating, but what are we to do if for instance we are called into rooms in the morning where there are contagious diseases, as diphtheria, etc. ? You know that it is not even doubted that under such conditions the safety lies in having full stomachs.

Signed,

MANY READERS.

This is a question no less interesting than important. When I was in attendance on those cases of diphtheria I saw them early in the mornings. *I took an empty stomach into the sick-room always,* when seeing them before noon, and for the following reasons :

1. Because I had no hunger in the morning.

2. Because I wanted the best possible use of brain function, where reason and judgment were wanted in my highest possibility, and hence,

3. I could not afford to part with any of its force to be exhausted over the disposal of an unbidden meal.

4. I could only lose by the mental and physical effort a little of those tissues which could be easily spared.

5. With the brain duly and adequately nourished, the blood and vital power would be in a stronger defensive condition against the millions of microbes that were doubtless drawn in with every inhalation. At this present moment I really know of no such potent defense against the microbe as *rich blood*. Had I taken an unbidden meal before starting out, I should have gotten well under way the first stage of indigestion, which by as much would have been the first stage of disease.

This would detract brain force from where, from professional reasons, it would be vitally needed, and would also take from the countenance something of that serenity of hope that needs to shine in the rooms of the dangerously sick, beyond all other places.

Again, I have often seen mothers watch over these cases day after day and night after night, and keep up their strength with their stomachs practically empty much of the time, and without getting a hint of the disease themselves. They could not eat because anxiety had created an intense aversion to food, and aversion is proof against all theories as to the need of filled stomachs in the presence of malignant diseases.

Now, my curious readers, it was safer for me to go into those sick-rooms than for you because of the absence of apprehension. It would have been all the more necessary for you to have gone there with empty stomachs, for the reasons you forgot when you raised the question. The shock of bad news and terror will paralyze digestion ; and fear, apprehension, depresses in its degree down to total absence of it. Shall you then fill your stomachs before entering such rooms for protective reasons only, fill up and then go into an atmosphere of gloom and of apprehension, where by as much the very draft that is the motive power of digestion is withdrawn ? Fill up the firebox of " 999," shut off the draft, and expect a mile to be made in *thirty-two seconds ?* I think not. You forgot not only all this, but that *wonderful bill of fare* which so cares for the brain, when you raised the question.

This is all the answer I am able to give, and until, if I am wrong, I get into the true light, I shall continue to recklessly expose my life, not only to the microbes of diphtheria, but to those of other communicable diseases by making my morning calls with an empty stomach, and making my later ones, after digestion has balanced the books for me.

And now for another question.

DEAR DOCTOR :—

There is a very prevalent theory that a human body needs the digestion of 59 oz. of food as a daily average. What have you to say to this ?

Signed,

MANY READERS.

To this I may reply that the theory is utterly absurd. As well supply such a theory to the firebox of a locomotive. As the size of the locomotive, the number of cars to be drawn, and the elevation of the grades determine the amount of fuel, so does it for every human being ; 59 oz. would be easily handled by the axman, but how about your professor who has had only that general muscular exercise called out by getting apparceled, and with the only addition thereto of a short walk to his class-room ? All day, and every day, his quantity of exercise will be infinitely less than the ax-man takes. How wisely has Nature provided for the needed exercise in the case of children whereby the appetite shall be strong and the ounces of food shall be duly disposed of even with less harm when in excess, than could be borne at mature age. It would seem that general exercise is very important in childhood, in boyhood, or there would not be those tastes that incite the rudest sports, the gymnastic tumblings, fence and tree-climbing, etc.

The number of ounces of food that any person shall take into his stomach during any day, must depend upon the exact amount of tissue burned up through bodily activity. As a rule the fuel offered to human waste is far in excess of the demand, and they are the greatest sinners in this way whose affairs do not call out general muscular activities, and the only way to meet this difficulty is to keep the hunger of child-hood and youth duly provided for by having the supplies offered at duly prolonged intervals.

And here is another very interesting question.

DEAR DOCTOR :—

Thus far we have not even once heard the word "bath-tub" mentioned ; if you have any views on this as a subject of any importance we should be pleased to hear them. "Cleanliness being next to Godliness," it would seem that so striking a virtue had not escaped your consideration.

Signed,
MANY READERS.

I will *consider* this question only. As a means of

"saving grace," the bath-tub is getting more and more into prominence, and what was once a luxury only to the wealthy, is becoming a necessity which is gradually getting into the reach of moderate means.

But the world at large must go on as the "great unwashed," and die prematurely, because lack of time and of means must avail against this "life-saving ordinance."

Mark Twain has described the pale, sallow, lifeless-appearing inhabitants of an ague-smitten district of the Mississippi who live, breathe, have a being, but who never move when it can be avoided. The shakes in their cases being merciful dispensations of Providence whereby they get needed exercise without any exertion put forth voluntarily.

The bath-tub as a modern institution is made the most of, as a general fact, by those who have the means and the leisure to use it for the alleviation of ailments that are never cured. I was once assured by a man in good business and social standing, living in a fine residence, with the bath-tub annex, who, beyond his sixty-fifth year, was the picture of robust health ; said he, "Doctor, I have not bathed in forty *years*, and my body is as clean as an infant's !" Was this a case of special "dispensation"? Or is there a general dispensation all along the "ranks" and "files" whereby life may be automatically preserved ?

Let us see : The human body, except the face, hands and feet, as instinctively shrinks from contact with water, especially if it be below blood heat, as does the body of a cat.

The human skin has a degree of oil in it which gives it due softness and prevents it from an undue evaporation of the water it contains. It also protects in a degree the escape of the heat of the body, as does the overcoat of blubber of the whale.

The human skin is covered with scales that are constantly maturing, and loaded with their odious accretions, are constantly falling off, each to uncover a bright new clean one, fresh from the hands of Nature.

Thus hath Providence provided that the " great unwashed " shall be clean. In this you will see that there is a line to be drawn between the flesh-brush and parboilings, daily perhaps, and the *forty-year plan*. The lack of bath-tub facilities scarcely appears in any work on the practice of medicine in a causative relation to disease.

And what are these accretions anyway, but unseemly aggregations of ultimate elements, chemically clean ?

It is my impression that many of you are spending too much of your time under water ; that it is unhealthy, unwholesome, unsanitary and disease inciting in the degree to which you carry it. True, you are able to apparel yourselves after those general soakings with a delicious sense of cleanliness, but there is a mental and muscular enervation for a time that points to the arm-chair, and not to the text-book or the workshop. You have denuded yourselves of those scales, and by the use of your alkaline soap have converted the oil of your skins into emulsions, and by as much *perspiratory* function has become disabled.

The bath-tub should be your servant and not your master ; and what am I to say of its use when the water is habitually below normal heat, and so used and habitually as a means to maintain health ? It seems to me that health which can only come, which can only be maintained by such habitual torturing, and time-sacrificing means, is scarcely worth the effort.

It is such a prime necessity to whale-existence that

his blood shall be kept habitually at the warm normal temperature, that he has been provided with an overcoat that permits him to push his ways along between icebergs without the loss of a degree of heat. But man, less protected, will denude himself of his scales, dissolve his thin overcoat and deliberately subject his blood to a protracted chill, that *chills mind and soul at the same time*, and under the delusion that he is wiser than nature! The general bath in proportion to its protraction, is a debilitating process, and all the more if in heat above the normal of the body or blood, and very much more if the temperature is below the normal.

How much you shall use your bath-tubs must depend upon your individual ideas of its need whether as a luxury or a necessity, and of the employments engaged in.

In a general way I can only suggest that you are to keep your bodies as clean as you can without impairing the structural condition of the skin by an undue excess in the use of the flesh-brush by which the scales are torn off before they have matured : that you are to avoid an undue use of soap by which the oil in the skin is converted into an emulsion, easily washed off. For the sake of health you need soft oily skins, not denuded, dry parchments. And you will not forget that, in proportion to the heat and duration of the general bath, so have you become disinclined to any mental or physical exertion ; and so will you need to have a care that you do not become unduly exposed to low temperatures with your thin overcoats left in the bath-tub in a state of solution.

The world's hard work is all done by those who know not the luxury of bath-tubs. Be clean, each one of you, as shall meet the individual sense of need, but go into your bath-tubs to have a good time, and not as the torturing doses go into the mouth, and down into the stomach. Go into those tubs with no more thought about your health than you have in your morning facial ablutions.

Life becomes a burden when time has to be spent over matters, sanitary in object, which do not involve either business or recreation in their execution, and all the more in proportion to the time and the regularity required.

In this connection I must speak of that refined, that heathenish cruelty, that strips the tender body of an infant of its clean clothes for a skin-denuding, oil-emulsing process. Enwrapped in clean clothes, kept in clean cribs, and yet its little tender body becomes so soiled during twenty-four hours that it must be soaked off, while the heat of the blood is being lowered by an atmosphere below the normal! There are a great many mothers within my circuit of influence who fully believe that once every week this process will do for the nurseries. And I cannot now recall a single instance where in a professional call to the nursery I have had to meet the reproachful look, the reproaching words, "We have not been bathing our child *often enough!*"

Readers, the soiled scales will continuously drop on reaching maturity; the skin has feeble absorbent powers; in a bath of laudanum we can hardly absorb enough laudanum to affect the brain; we cannot get local anæsthesia from cocaine without sending it beneath the skin through a needle. The soiled scales are liable to become absorbed. But, dropping as the leaves in the forest when their fullness of time has come, they become entangled in the clothes; is it then just barely possible that these clothes have more need

to be changed frequently than is needed the soakage
of the general bath ?

Readers, draw the line upon the bath-tub anywhere
between the extremes I have given you that will
permit its use as your servant, and not your master.
Use it with the same idea of its relative importance
as you use your wash-bowl ; use it with that same
indifference to its sanitary importance. Life is too
short, there is too much to do, too much to enjoy, to
have time needlessly wasted in any worrying means
and thought about what you shall or shall not do to
maintain health. What you should seek is the child-
like indifference to all such doings and worryings.
This can only be done when life-habits have been so
cultivated, so adjusted that they can go on naturally,
physiologically, automatically.

Next.

DEAR DOCTOR :—

Your statement that the weight can be so reduced
as to meet the ideal in actual pounds seems almost incredible, very hard indeed to seize upon with any
force. Can this actually be done with no involvment of
danger to health or to life ? Something still more on
this subject will be of interest.

Yours,

THE HEAVILY LADEN.

In reply let me again call your attention to the fact
that escaped your memories when you asked the
question.

If the actually sick can go on day after day losing
weight, and anywhere from *one day to sixty or seventy
days* with so little danger to life that the greatest
majority of cases recover, why cannot the brain care
for itself equally well when an artificial fast shall be
entered upon that involves no brain taxing ?

These fasts have been entered upon for various reasons with sufficient frequency to indicate their safety, even when protracted a month or some days beyond a month. I would have you all understand that as soon as the fast begins the loss of weight begins, and the rapidity with which it shall go on, will depend upon the expenditure through exercise. If none is taken, the loss during each twenty-four hours will be very nearly the same, and in such a case the scales will enable you to very nearly indicate what your weight shall be on any special day or time to come. You can reasonably expect to remain alive until your weight has become reduced to very much below the normal, and I am unable to see any reason why not with perfect safety to health. If this excess has largely come to you through exceeding your need of food, it ought to be safe to reverse the process.

This idea fully takes into account that the normal weight of the individual, includes a great deal more of the adipose tissue in some than it does in others. What you all have need to do, is to work off the surplus which greatly increases your weight above the normal. And the way this is to be done needs no further hints from me.

My next talk will be to some of my railroad friends.

LECTURE XXII.

Gentlemen of the Iron Horse :—

I have invited a few of you to meet me to hear a short talk about the way that you shall care for your stomachs, that you shall get the highest available functional power of your brains. There is no business that men ever engaged in that so involves the safety of human life as does yours. No business where so many lives must be held for hours in the hollow of a single hand.

I meet you with all the more interest because I know something about your life-work in a practical sense. I too was once a trainman. Wherever any one of you has ridden on a train except upon the brake beams, I have ridden. And I have ridden over poorer track than most of you ever saw. It was my fortune for a time to be a surgeon in charge of a train of hospital cars, that in war-time gave me all the danger I might expect on the skirmish line, large experience in getting ditched, and for variety, an occasional volley from strolling "bushwhackers." And I have ridden hundreds of miles, when my life was in the hands of an engineer who never got on to his engine without his alcoholic dose, who never got off without an immediate alcoholic dose, and who never knew one moment of comfort except when under the influence of doses. And I am alive to be before you to-day !

There is nothing of broken rests, days of duty not

271

spaced by nights of sleep that I do not know all about. There may be some yet living who will recall what train-life was, when for a month it was our duty day and night to relieve the army of Gen. Sherman of his disabled that he might cut loose for the sea.

That was train service, gentlemen, that had every element of exhausting service, incident, excitement and danger that might be coveted by the most reckless, or the most patriotic.

Napoleon is said to have lost a battle because of an attack of indigestion due to an unbidden meal. Every faculty of that vast brain had become dulled, and by as much his splendid army was disarmed and hence defeated. How many switches have been left open, how many train orders have failed to impress or have been forgotten, because of dulled intellects, dulled memories, by reason of attacks of digestive torpor, belongs to the unrevealed history of tragedy.

Gentlemen, you can only reach your highest mental possibilities when your stomachs are empty, and there is no division so long, that that engineer is not better equipped in memory, in judgment, in every possible mental faculty and in physical endurance to guide his train over every mile of it with his stomach empty, than with the least food in it.

This you think is a remarkable statement, and since you have had no time to attend my lectures I shall give you a little outline of theory or rather of digestive facts, just enough so that you can grasp the idea.

Digestion is just this much of a tax on brain force that the mind is dull while it is going on. You go on to your engines loaded; there has to be a division between the working forces of the stomach and the working forces of the mind, of that nerve or brain force that either wants in full, for perfect work.

As soon as you get on duty one or the other of the
working parties is bound to suffer, and it is both.
You worry through, you who are the strongest will
get through with the most ease.

Is this an ideal sketch ? I have been told by some
among you of an occasional nod while on duty, and how
easily a station or a side-track might have been passed
and a meeting of trains the result. That nod, that
fitful sleep, never would have occurred but that a weak
stomach was exhausting brain power by a great effort
to dispose of a meal without digesting it.

Gentlemen, I am going to ask you to accept the statement that you can run your trains over the longest of
your divisions and reach the end of the last mile in a
stronger condition in muscle and mind with the stomach
kept at perfect rest, than can be when it has been
engaged in a laborious effort to perform its function,
with very much less nerve stimulus than is the imperative need ; I am going to ask you to accept this statement as a fact until you become easily convinced by
experimental evidence.

Now on arriving at the end of your division, I am
going to ask you to accept this statement as a fact,
that in exact proportion to the muscle and mind
exhaustion, in other words, in proportion as you
are tired, so will also be your stomach, and so it
has become disabled from performing that service you
have habitually called upon it to perform regardless of
all question of its condition.

Let me illustrate : A few years ago one of your
number, then a fireman, came to me one evening with
the story of terrible headaches. On the previous night
he had had the worst attack he ever had in all his life.
Now he was a giant in strength and of almost perfect
physical development and proportion.

18

Said I, " My dear sir, you ought never to have a headache, what have you been doing to invite it?

" Nothing that I am aware of."

" When did you go on duty yesterday and where ?"

" At 7 A.M. in the morning, and fired a switch-engine all day."

" Did you eat a hearty breakfast ?"

" Well, yes, rather."

" Were you hungry when you ate ?"

" Well, hardly."

" Why did you eat then if not hungry ? "

" Why, to keep up my strength until noon."

" Did you feel refreshed after this breakfast ? "

" I can't say that I did."

" Have you ever noticed that when you have become real hungry that not only you enjoyed your food very much more, but it seems to make a new man of you even before you get up from the table ? "

" Why, certainly, that is always the experience."

" Now let me ask you if it is not likely to be a fact that putting that breakfast into your stomach when you did was not to forestall a case of hunger in mid-forenoon ? "

" Why—it looks that way—why, yes, certainly."

" Well, now, would it not have been just as well for you to have carried that breakfast along in your *tin* lunch-pail until a need was created ? "

" Why—yes, certainly, but you know I would be on duty later and could not well stop to eat."

" Really, now, let us see about that; do you undertake to say that in a case of real ravishing hunger in mid-forenoon, you could not have found time to relieve something of the urgency between your shovelings ?

" Remember now, that in real hunger, a crust relishes much better than the nicest spread in the morning, when you are not hungry, and by as much it would refresh you immediately."

" Why, Doctor, I never thought of this in this light before, and I easily see that I converted my stomach into a lunch-pail before starting out; and I remember, now that you have presented the matter in this new way, that my breakfast did not prevent a feeling of exhaustion during the forenoon, and could I have been relieved from duty then I should have sorely wanted a hearty meal."

" Well, what about dinner ?"

" I took it from my lunch-pail on the engine."

" Why on the engine ?"

" Because I had no time to leave it for a general meal."

" Did you relish it ?"

" Very much more than I did my morning meal."

" When did you get off duty ?"

" So as to get home at 7:30 in the evening."

" Were you hungry then ?"

"Indeed I was."

" And tired ?"

"Very."

" It was almost hard work to breathe, and when you got down into your arm-chair you felt too tired to even talk, indeed every muscle of your body seemed to be ' fagged out.'"

"That is a very accurate description of the condition I was in last night when I dropped down into my arm-chair, Doctor."

" When did you have your supper ?"

" At about 8:10."

" What did you have ?"

"Well, let me see—I had boiled cabbage, beans, pork steak, etc." (this is no ideal sketch).

"Did you eat heartily ?"

"Indeed I did."

"What time did you go to bed ?"

"At 9 o'clock."

"Now you are very certain that your arms, your shoulders, and in fact that all the muscles called into use by those long hours of shoveling were actually tired, are you ?"

"Why, yes, certainly."

"And that even your mind was tired as well ?"

"Certainly."

"Well, now, how does it come that there should be a strong, a vigorous, a lively stomach in your tired body which would be able for such a task ? What kind of a man are you anyway that you could expect to have such a conditioned stomach after such a day's work, how could you think of going to bed to rest all other muscles and have this one muscle, almost the most important, perhaps the first in importance, in your body, to worry through the livelong night, over-taxed, over-powered, enfeebled, as tired as any other muscle of your body ? Was that a square deal ?"

"Why, really, Doctor, I certainly never thought of this in such a light before, and I never shall forget one word of what you have told me."

"What time were you called to account for your sins against your stomach, in other words, when did the 'strike' come on ?"

"What do you mean, Doctor ?"

"I mean, when did your head begin to feel the splitting head-pressure ?"

"What, am I to understand that my headache last night was due to that supper ?"

"That is just what you are to understand and nothing less."

"Well, well—I only supposed that it was an aching such as I have been subject to for years, only for some reason, not clear to me, was worse than usual. I begin to see—yes, my eyes are getting open to a new light."

"Well, my dear man, you have called upon me to get something to relieve those headaches. What particular drug will you prefer?"

"Why, Doctor, indeed, I think if you will excuse me, I will take some warnings from your talk and let the drugs go for the present. No, Doctor, I think I will take no remedies home with me to-night. But you really think that all my attacks of headache have 'come from my stomach'?"

"Certainly, every one of them from the abuse of your stomach in the way I have outlined."

"Doctor, I am very glad I called, and since you have presented this matter so clearly, I shall go to my home with a feeling that I need have no more headaches, if I am duly careful."

"That is just the way you should feel on your way home, and that thought ought to be the first of every morning, that your life shall be guided into more healthful ways."

Gentlemen, that man was wise enough to comprehend and to be abidingly influenced by my lecture, and not only did he relieve himself from those attacks, but he has been infinitely safer as an engineer ever since.

Two years ago there came to me an engineer of many years of service, but in the very prime of life. He was unduly weighted by bloat and excess of fatty tissue. His was a complaining stomach. He was a

sober and industrious man, a man who cared for his
family. He had no bad habits, no expensive ones ; was
always found ready for duty when not disabled by his
stomach. There was very little of cheer ever seen in
his eyes or upon his weighted, hanging face.

He was always doing more or less to switch that un-
ruly, that unreasonable stomach on to the main track.
He had pushed his efforts in this way for many years,
and yet it would persist in remaining " side-tracked."

I gave him an address, and he left to do some thinking. The first time he reached home after a "round
trip" he surprised his wife by calling upon her for only
a cup of broth before going to his bed ; perhaps she was
worried, because, after the long tiresome trip, that he
would think of going to bed without the hearty meal
he seemed so much to need. But she was not used
to posing any questions at him as to what he should
or should not put in his stomach, nor as to when nor
how or even what. And not only was she surprised
at the continuation of this new eccentricity, but was
further surprised that whenever called from his sleep
to go upon duty she had only to get him the least of
all possible breakfasts, and still more that there was an
ignoring of the lunch-pail also. What did it all
mean? She did not care to investigate ; she had not
the remotest idea of the origin of the whim until
months later. But there was a greater surprise in
development going on right under her eyes. The
home seemed to be gaining in cheer at every "round
trip." Later on I heard this from the wife. "Doctor,
I have no language to express the change that has
taken place in my husband ; to understand it as it
now is, you would need to have actually lived with
him, before, and for a time since he began this way of
living to comprehend it fully."

Gentlemen, that same change by as much has strengthened his memory and his judgment, and he is not only infinitely happier than he was before, but by as much, his services are of more value to his company. Richer blood means better judgment, a sharper memory and a higher sense of duty to his employer!

This engineer is losing no more " round trips " because of physical disability, and he knows that his human fire-box must be used with something of the same idea of fuel in proportion to the demand, as he requires of his fireman. He will not load up, fill in, and close the draught any more while on a side-track. And he will let his fireman have an easier time when on "down grade."

Many of you are heavily-weighted ; let me ask you to accept it as a fact that all this is surplus food that will duly nourish your brain as your watchful eye is keenly directed ahead. That vision will be all the clearer, the senses all the sharper as you hold those hundreds of lives in your hands, if you have not disabled yourself by filling your fire-box and closing your draughts.

Gentlemen, I beseech you, take those hearty meals only when you are rested, and have time and leisure to eat and digest them, to use your lunch-pails only in cases of emergency, and never again presume to enter upon any rest without treating your stomach with that care its transcendent importance indicates. Be *reasonable and impartial.* Let me urge upon you to treat your stomach thus justly during a few trips, during the intervals between the trips, and you will get so overwhelmed with conviction that nothing of pecuniary estimate will induce you to go back to your old ways, which you will even know as suicidal.

Make it a supreme law of every day of your life

that you shall be able to have at least one full meal when you are *hungry*, when you are *rested*, and when you are to remain long enough in a state of rest to be able to convert it into living tissues.

In the irregular way you are compelled to perform your duties it is often difficult to do this, but always aim to do the best—not forgetting that your only danger can come from eating too much, and never too little when you are holding human lives in your hands.

And what is needed in the " cab " is no less needed in the despatcher's office.   Only in the unveiled history of tragedy is it recorded, how many times your comrades have gone down to death because digestive tax in the stomach and the resulting " muddled brain " had sent the fatal order to go ahead on the *main* track when the side-track was meant.

Begin at once and there shall come brighter skies for you to gaze upon, brighter homes for you to live in ; and infinitely safer engineers to be held responsible for human life inestimable, and for property of vast value.   Take your meals when rested only, take them never without hunger, and you will rarely miss a trip because of sickness, and you will live for more years of better service in every line of your human duties.

LECTURE XXIII.

Gentlemen of the School Board :—

I have invited you to meet me this morning that I may give vent to certain swelling emotions on the subject of the preliminary education of childhood and youth.

" We must educate, we must educate," we used to read on one of the pages of our McGuffy's Fourth, and then how ingrained becomes that trite couplet :

> " 'Tis education that forms the common mind ;
> Just as the twig is bent the tree's inclined."

It would seem that upon that couplet as a basis has been built our modern school system.

Gentlemen, the educative machinery of our city in its comprehensiveness, in its exact adaptation of means to ends, even in the minutest detail of supposed needful means to desired ends, is as perfect, as educative machinery can well be, and it wants nothing of motive power in the execution of its processes.

There are eight hundred boys in the various ward schools in daily attendance and perhaps as many girls, but for the purpose of directness of thought I will talk about the boys.

You are proud of our city schools and of our school system, and of what is being done, that those boys, those tender twigs, shall not become the bent, the twisted trees in manhood's estate ; there are eight hundred of them in the ward schools ; enough to make nearly a regiment of soldiers.

Gentlemen, your beaming, satisfied faces are often seen among younger and happier faces in the assembly room of the high school, on the long-looked-for and most-ardently-desired graduating day. You have listened to two score of short pithy graduating essays, read by pale-faced young misses ; and you have heard two, three, rarely five orations from youths whose faces are not quite so pale, five out of eight hundred. Are you not proud of such a triumph of your educational plant ? Are those not a very few kernels to be threshed out of so very large a sheaf ? Is not that a very small " grist " for so many years of " grinding ?" Where are the "ninety and nine" twigs that were being educated in such large numbers to become the symmetrical trees of our adult life ? They are where you will always find them until our school-law is enlarged.

Why is all this ? Gentlemen, those rattling boys are in the ward schools by virtue of the right arms of their parents. They are spending six hours of five days in every week, and during nine months of every year, in crowded rooms, inhaling air heavy with impurity, air devitalized of its oxygen, and every breath so drawn is a violation of constitutional law, of law that must be obeyed if man's estate is to be reached in highest fruition.

Gentlemen, parents can have no higher duties in life, no more serious cares than those involved in keeping up the highest possible physical and moral culture, that their sons, that their daughters shall reach mature age in the most perfect physical and moral condition. Your sons have so come into this world that for the first fourteen or fifteen years of life they need to be happy every moment, that the nutrition of their bodies shall be kept at a balance. You cannot put one of your boys at a task without an instant diminishing of this life-breeze which is their right as is the very oxygen of the air. Every hour of irksome confinement is a disease-inciting ordeal.

Many of those boys are the hapless victims of weaknesses due to heredity which, as unseen foes, are awaiting the due amount of culture when they will enter upon their harvest of misery and death, and there will be a wondering over the mystery of "Providential ways."

Gentlemen, a few years ago there was a family of two sons and a daughter that had all the care that devotion could provide. The boys died at an early age of obscure brain disease. When the daughter had reached the age of eleven, I was called to see if I could so build her up that she could avoid the frequent sick-spells that were seriously interfering with high percentage.

I found hints of brain trouble, and I spent an hour pleading for the child's life, and the result was that it was decided that a daughter with no further school-culture would be a great deal better than no daughter. It was advised that the child be permitted to grow up with the largest possible recreative privileges, and with never a hint that anything in print, not recreative in its effect upon the mind, should ever be even suggested to her.

She is a young lady now, and since she has had no need to teach for her daily bread, she seems, in her self-assurance, in her command of language, in the perfection with which she can perform all the duties of young womanhood, about as well equipped for the work her brains find to do, as if there were vague memories of what was found in sundry text-books.

The boys remain the prisoners of the ward schools until their sturdy right arms get too strong for the parental, and then the faltering by the way, the dropping out begins ; and the war against constitutional law, as conducted in the school-room, ends for them ; or if there is to be more culture, it will begin when developed tastes will make the text-book a recreation. Is not this a fact, gentlemen, beyond any question ? Go where you will, ask any educated man when his culture began, and he will very generally place it at that age when the city schools end it ; he will give the twelve or fourteen years of preliminary educating attrition very little credit for its ability to enable him to comprehend anything in text-books that appeals to his tastes or, what is the same thing, his recreative sense.

What is the ward school doing for the boys anyway ? I am not going to attempt any answer. I will simply suggest things that I think education does not do. I may assert that it cannot determine what any human employment is to be. The tastes will determine all that in due time, and according to the mysterious law of "natural selection."

Moral power it seems to me is a matter of heredity very largely. Men in business who have need of day-books fully realize this, and they have very little faith that moral culture has much to do with ardent desires for balanced accounts.

What is right and not right between man and man seems to be determined by a natural sense of justice or injustice, and not particularly because there is a twentieth chapter of Exodus. Your boys, gentlemen, are to be carpenters or printers, clergymen or farmers, missionaries or perhaps bank-robbers by reason of the irresistible law of " natural selection." And the ward schools can have little power to determine that the selection shall be worldly wise.

Again I ask you, what are the schools doing for those boys ? No one of you will read a book or even an article in a paper unless it interests you. At your mature age you would think it utter folly to undertake to master any study for the sake of culture alone, particularly if the subject of it were void of all interest. How have our college chairs been filled ? All by men who followed out their line of tastes after due age had developed them. There had to be keen interest on every page or it would not have been mastered.

Your boys are expected to get high marks from poring over text-books that are void of every human interest, and when every instinct is alive with the desire for the fresh air, for recreative joys. What do they care for high marks ? What other care is there but high marks as a matter of pride and not as a matter of vital knowledge or of culture ?

Travel the world over, gentlemen, and ask every learned or any business man the secret of his learning, the secret of his success, and he will never go back to the school-room where he felt himself a prisoner for six weary hours of the five out of the seven days of the week, as the time of high resolutions to achieve a name worthy of honor.

It is my opinion, gentlemen, that your vast machinery is expended mainly on the memory. Now memory, as a faculty, is held to little account for life's successes or life's failures in this world, except in the cases of trainmen and switchmen. Those lessons are generally memorized for the express purpose of meeting the demands of the marking system, and not for the culture there is in them.

In this life we seldom forget what is vital to our happiness. "I shall never forget," said Artemus Ward in speaking of one of his friends, "I shall never forget how he used to borrow money of me!" And so do we all remember, and as we have the need whenever anything vital to our happiness is in question.

Now, gentlemen, I am not going to advise you to abolish the ward schools, but I am going to suggest some radical changes.

There are very many delicate children who have to take long walks to reach the places of their confinement. They get up and swallow hastily their enforced breakfasts and then rush out for their journey. At noon the walk has to be doubled, because they are not permitted their lunch-pails. Their dinners are also hastily eaten, that a little time may be saved for recreation in the school-yard, and it is only a very little they can get in this way. As a result there are dull intellects during three hours, low marks and disheartened teachers. And this goes on day after day, and the result is a steady going down of the health of very many of the weaker ones. It has very often occurred to me to advise parents to give up all idea of putting their delicate children through those long years of constitutional, taxing experience.

Now, since you are not able to get more than the three or four boys into the graduating class each year, three or four out of the eight hundred, and since you cannot force them to study or be in love with learning, and since, in spite of all your efforts, they are going out into life to drift or to push their ways along as moved by the motive power of their human interests, I am going to advise you to do this, or, in order to express myself with more ease, I will tell you what I would do if I were the whole school-board myself, and invested with autocratic powers.

I would make the morning session two and a half hours long, and have it begin at eight o'clock sharp ; I would have every pupil appear after a breakfast of nourishing broth only. At 10:30 they would be released, to go leisurely home, getting all they could out of social recreations on the way ; then in due time they should have their hearty dinners—and then until 2:30 to exercise them into new flesh and blood. I would enlarge the school-yards and encourage recreative sports. Did it ever occur to you that nature has not been unmindful of the mind while the body is in a state of development ; that in the school-yard there is more real healthy exercise of the mental forces than is ever brought out by odious text-books ? Do not their little games call out every power of mind in instant action, and action strictly in line with the tastes ? Did it ever occur to you that there was mental culture combined with physical culture in the noisy commotions of the school-yards during recess ? That it is this clashing of mind with mind in recreative rivalry that makes those minds as well as bodies stronger ?

There is a culture in the streets of the great cities, and for instant and powerful use of the mental faculties in all the lines of self-interest the bootblack, the newsboy towers away above his cultured fellows of the school-room, age for age.

I would have the terribly exhausting work of the teachers cut down to a living rate. There is no human employment, that I am aware of, where culture is wanted, that is so exhaustive of vital power. Your professor in his easy-chair can have an enjoyable life in teaching his favorite science ; it is recreation to him, for there is always something new to add the charm of interest to his duties.

Not so for the teacher of the primary department; it is the same simple knowledge to be taught over and over, and week in and week out. There is no stimulus of novelty, of self-culture in it. Those teachers are the hardest-worked of all public servants ; and I would so cut their work down that they might meet their pupils with faces beaming with the brightness of health every morning. I would abolish those careworn faces, eliminate the rigid, drawn lines, and I would do this by allowing a balancing of the books after the last session. I would have the exhaustion abolished by rest. There should be no tiresome exercises to go over after school-hours, whereby sleep becomes imperfect and digestive action becomes enfeebled.

I would eliminate all taxing studies, especially those that do not appeal, in the slightest degree, to the re-creative sense. I would abolish the dead languages and substitute for them more culture in the written, the spoken use of the parental tongues, the living languages of to-day.

The dead can be raised, will be raised later on when natural, self-generated tastes shall furnish the motive power. Your crowded school-rooms and enforced studies do not determine who shall fill the vacated chairs in the University. Nature herself is the determining factor in all such questions.

But eliminate as you will, add as you may even to the utmost to the recreative element of school-room attrition, make the sessions as short and enticing even as the Sunday school, there will be the inevitable outcome, eight hundred boys in the class-rooms with only one, two, three, four or five to reach the graduating class because yours is a way against the forces of growth of development, and you will never reach a larger per cent. until the iron hand of law becomes a re-inforcing power, and because those long hours in rooms, now too hot, next too cold, in rooms with air heavy with impurity, are hours of diminished draught to all the motive forces of life.

You can do a great deal to lessen the force, the attrition of this war against the Heaven-ordained rights of the growing boys and girls, and it will have to come by adding to the recreative needs in all school work. I tell you, gentlemen, a system of school work that cannot be carried on without such destructive results upon the health of the weaker children, and upon all of the weaker teachers, is not one thing less than *barbarous, because it is death by the slow, the inhuman process of torture,* and it will be so regarded in the time to come. I ask you to go to the schools and see the careworn expression upon the faces of those teachers, the outer signal of the inner distress, look at them and remember that their six hours of labor is so exhaustive that Nature ought to have eighteen hours in which to balance her heavily overdrawn accounts.

Were those frequent overpowerings of teachers and the weaker children to be thought due to bad sanitary conditions of your building? How swift you would be to relieve the causes!

Gentlemen, on this particular subject I feel that I know whereof I speak, for I have for many years been brought in direct contact with the results of your exacting system in dealing with disabled teachers, with disabled children; and what may I not expect when law shall make it a crime not to send those fresh air plants to your crowded rooms, out of the sunshine, out from their vitally needed life breezes, to *wilt* for six long mortal hours, only to harass the teachers by an

19

utter indifference to the vast importance of high marks. High marks, gentlemen, often mean nothing more than good memories ; they bear no assurance that in the practical affairs of life they will have a shade of determining power as to success. In your own affairs, gentlemen, you are pleased to think, to believe that your successes have all come from a wise use, not of your schooled, but your *self-cultured, natural* powers. There is a bit of self-glorification in the thought that Nature was kind enough to be a little generous in gifts of natural endowment in *your cases*, and not in the least will you remember those care-worn teachers who worried over you for so many months out of a great many years, as deserving any credit for those successes, for these have all come through *special dispensations* in the way of your *natural gifts*.

Gentlemen, it is within your power to so reorganize your machinery, as to largely relieve it from its destructive effect upon the health of growing childhood and youth. It is your duty to so reorganize it that the feeblest boy or girl can approach the average degree of vacation health ; to so reorganize it that the feeblest teachers can appear before those young lives as ideal women, every morning of the long sessions, because they have had time to eat, time to digest, time to sleep, time to become regenerated.

A countenance beaming with a glow of soul has infinitely more power of appeal when directed to those uneasy chairs. Those teachers no less than the pupils, gentlemen, have a Heaven-ordained right to have their work so adjusted that the highest possible physical condition shall be maintained automatically.

Did you ever read the following epitaph to a teacher?

> " Among the many lives that I have known,
> None I remember more serene and sweet,

More rounded in itself and more complete,
Than his, who lies beneath this funeral stone.
These pines that murmur in low monotone,
These walks frequented by scholastic feet,
Were all his world; but in this calm retreat
For him the teacher's chair became a throne.
With fond affection memory loves to dwell
On the old days when his example made
A pastime of the toil of tongue and pen;
And now amid the groves he loved so well,
That nought could lure him from their grateful shade,
He sleeps, but wakes elsewhere, for God hath said Amen."

Thus wrote Longfellow, the pupil of Parker Cleaveland, the teacher. Ideal teacher, ideal pupil! There was a pupil old enough to have study made into a pastime, a teacher who, by reason of having his favorite science to teach, became a king among teachers.

The text-books of your ward-schools, gentlemen, are in dead languages to the teachers; they go over them term after term, and there is no recreative force in a single line. And the sixth hour of each day is reached with a feeling that there has been nothing of self-culture added to the autocrat's chair, and only per cents among the seats painfully low!

Gentlemen, I have one son in his thirteenth year who is attending the high school. I am pleased to have him there for what he gets of discipline in the matter of reverence for rightful authority over him. I am glad for that culture he gets in mathematics and for all that he gets that will enable him to speak and write the thought within him with more ease and correctness. I do not expect that his schooling will tend to enlarge the fountain of thought, any more than I can expect to have powers poetic planted and enlarged by such culture. I am pleased to have him under teachers who are well worthy to be moral guardians no

less than intellectual.   I am pleased that he should have
that large opportunity to measure his mind with so
many other minds in the rivalries of social, recreative
clashing : a culture that must go on with him during
all his life.   I have little care as to high marks because I know somewhat of his natural gifts.   My
chiefest care is that there shall be first physical and
moral culture and second intellectual culture.

Not one of those teachers has any need to lose one
moment of "blessed rest, of blessed sleep," during
regenerating hours because I am ambitious for high
marks.

This son goes to the school-room every morning with
an empty stomach, hence he has the highest possible
reach during all the forenoon in mental conception and
grasp.   At noon he appears for his general dinner with
no hint of exhaustion in manner, with nothing of the
look of the released prisoner.   He returns to his three
hours of confinement with a duller intellect because
your school-engine has started a full hour and a half
before he could recreate that dinner into life force.
But he is better able to meet the adverse conditions
than his fellows, because a rested stomach is a strong
one.

Thus is he meeting the demands of school culture.
And when that legislature shall pass a law that makes
me a criminal if I do not enforce his attendance after
it shall become in the least evident that fresh air and
freedom is a greater need, I shall be pleased to set fire
to a train that shall reach a chemical combination,
which will make the erection of a new state capitol an
immediate necessity.

He will continue his attendance until, through the
evolution of natural law, the text-book and the long
hours have become odious, and by as much paralyzing

to life force, and then it will be a stronger right arm than mine that shall enforce continued attendance.

Such, gentlemen, are my views of preliminary education as I find it in operation in the ward and high schools, and during those years when nature is continuously crying aloud against its enforcement.

> "Hark! from the tombs a doleful sound,
>   Mine ears attend the cry!"

Gentlemen, cultivate your sense of hearing, your sense of understanding, that all the audible, all the warning voices of nature may be instantly heard and understood, and duly heeded.

LECTURE XXIV.

IMPORTANT QUESTIONS CONSIDERED.

INSANITY—THE GERM THEORY OF DISEASE—BRAIN WORKERS—SO-
CALLED MEDICINAL WATERS—FRUIT—BILL OF FARE FOR THE
HUNGRY—MORBID HUNGER.

My Friends the Readers :—

I find that I am to spend the time of this morning's
meeting in a consideration of certain questions you
have raised, and the first one is of exceeding gravity.
I can only consider it, because my experience has been
too small to illustrate with that force I would like to
be able to possess when confronted with the consideration of so grave a disease as insanity. The question
is as follows :

DOCTOR,—You have told us of various diseases that
seem to be easily cured on the physiological plan of
life, but so far you have said nothing as to insanity.
Are we to infer by your silence that this is an exception to all other diseases ?

(Signed)
MANY PAINFULLY INTERESTED READERS.

It is a striking coincidence that I was engaged in
some thought over this question, when there chanced
to come to my home by the sea a physician from a
large city attended by a nurse; the physician was
a specialist in the treatment of the diseases of the
nervous system, but as it was alleged, had gotten so

worn out by over-work that his own nervous system had become so demoralized as to make special attendance on him a necessity. I noticed him with some care as he presented himself to his host to be assigned to his room, and there was a sallow complexion and a somber expression. There seemed to be no life or cheer within.

His nurse took him to his room, and then, as he was about to go to his bed, the nurse had his own bill of fare taken up to him—which his doctor patient was expected to eat, and because it was of most excellent quality for any ordinary case of hunger.

The patient went quietly to his bed after an interview with a physician who had been called to advise, and to whom he spoke about a recovery. But in the morning he was found with his neck attached to the bedpost and his disease was cured. There was to be no more trouble with his brain.

Now in this case there were behind the mental agony weeks, months and years of brain taxing due to the exhaustion over the disposal of food without digestion, this agony being so intense that he could strangle himself with no sense of physical pain, for he had to flex his knees in order to get the weight of his body as the breath-stopping force. And hence his constitutional tendency had been thereby duly cultivated to this tragical end. Such I believe to be the case.

A few years ago a man fell into my hands who was in attendance upon all his affairs, but yet to his intimate friends had been revealed symptoms of mental disturbance of a very serious character, and they were associated with marked dyspeptic symptoms. His eating habits were gluttonous in the extreme. I succeeded in getting a radical change in a state of evolution in the mental condition.   At this time that man is not only in appearance perfectly well mentally, but I believe he will remain so while he keeps to his better living habits.

One of my fellow-citizens went to a hospital for the insane a few years ago after a deed of violence that made him a danger to his friends.   He was a man past the middle period of life, tall, erect, of free, easy movement, amiable and of full courage in moral and mental force.   By reason of his business and of late hours he became the victim of some drug-habit which, with the irregular sleep and the habitually paralyzed digestion from his drug, and the exhaustion resulting, made him dangerously insane.

A few months after his incarceration I saw him and the well-built form had become weighted with not less than seventy-five pounds of fat and watery tissues, and later he died.

On getting to his new home, the system being relieved from the taxing of the doses, he began to eat the usual three times a day.   Now as he took much less exercise than before, and was compelled to be in bed many more hours on the average than for years, the surplus food began to weigh him down in the form of fat, and then later on the overtaxed stomach failed in its work ; the blood began to get thin ; the water began to leak into his tissues, and the way was prepared for the supervention of some acute disease which took him off.

Now in this case, and in all such cases and in particular where there can be little exercise taken, I should keep all on substantial meals late in the morning, and light second meals at least four hours before bed-time.

I have told you that it is my conception that there is always a physical basis for insanity in a structural weakness due to heredity. This only awaits a due degree of development by avoidable or other cause when the disease shall become manifest in its multiform symptoms.

Children are not born insane : insanity is rare in childhood and youth, hence I conceive that the insane possibility existing in a structural weakness has its development in precisely the same way as do all other diseases local or general, and that its cure involves the evolutionary method in reverse as do these diseases.

Our hospitals for the insane are striking illustrations of the upward march of civilization. In their appeal to nature in disease they are far in advance of other hospitals. This is to be seen in the nicely-fitted rooms, the regular times for sleep, for exercise, for meals and in the means used to divert and cheer the mind. It is a most striking advantage that the wants of the insane, whether mild or furious in character, are always met with force kindly, strongly administered— it is of a most striking advantage that these wants are met by those whose bodies are so rested that there can be the highest possible degree of self-command, under provocation. If all these advantages could be supplemented by eating habits more nearly gauged to the waste of the tissues as caused by exercise, if all were to be fed not by the time of day but by the time of need, it is my opinion that the per cent. of cures would be largely increased. You easily see that I have some right to assume this from the almost instantly perceptible improvement in the mental condition of all patients who aid nature in this particular way.

Now comes a question as to my opinion of the germ-theory of disease asked as follows :

Doctor,—You have frequently alluded to the germ-theory in a way that does seem to indicate that you are strongly impressed with its importance, hence we would like something more definite from you on this subject.

(Signed)

MEDICUS.

In reply I shall have to confess that I have been finding so much of disease that is not only avoidable and also relievable that I have not given this matter such attention as to entitle me to an opinion.

It has been my fortune to have been so often condemned because of strong opinions held without "shadow of reason" that I have all the more felt the necessity of not having opinions without reasons to support them.

It is my impression that disease has its origin and development before the germ disturbance can become possible. In other words, I may assume that normal health and its rich blood makes the germ disease impossible. This is the "strong-man-armed" condition of the body wherein lies the true defense against all causes.

But I have had an impression that, as the human body is the natural place for many different forms of germs, they may act as scavengers to neutralize degenerative products by their passage through our bodies, subject to forces in vital chemistry.

In a former lecture I told you that the entire medical world was concentrated upon a possible specific for the germ poison of diphtheria, and now I am able to tell you that there has broken into the listening ear already one echo.

A medical journal of a great city informs us that the remedy has been applied in three cases with only one death, and hence the treatment has come to stay! "Reductio ad absurdum."

In a western town of the same state during the past two months, the local prints announce that there were thirty-seven cases of diphtheria with only seven deaths. Now these were subjected to treatments multiform. Had this new germ-destroyer been used, what effusion of ink would there have been!

I shall not be able to get any enthusiasm of expectation over the possibilities of this new means of attack. I am glad that there are so many of larger faith to go on with original investigation, and no one will be more ready to adopt any results that shall come which will advance medical science than myself.

And now for a very important question.

Doctor,—There can be no doubt we think that your idea of awaiting upon due hunger before the first meal is taken is essentially sound; but is it well for bankers, merchants and others who have a great deal for their minds, and very little for their bodies to do, to eat a hearty meal during business hours, and by as much get into sluggish mental condition?

(Signed)

Many Readers.

No, it is not. With the distinct understanding that any banker, merchant or others can do better mental and physical work with stomachs absolutely empty until hunger comes you can have no doubt that the breakfast, even if the slightest, will be by as much a clogging tax upon brain work. To any one a forenoon of the best brain work is assured with the stomach empty. If there is to be an afternoon of brain work there must be very little of a diverting tax from the region of the stomach. When the time comes for relief from mental strain, so that digestive force can have all the nerve force needed, then is the time to take that heavier meal ; not only with keener relish, but with the added pleasure that comes from the ending of the toils of the day. Farmers and all engaged in heavy labors need the first meal of the day, before undue physical exhaustion has been reached, because this is digestive exhaustion as well. There is always a very tired stomach in a very tired body, and the fact cannot be too carefully taken into account in all physically taxing employments.

I wish you all to most distinctly understand that morning hunger after a night of sleep is nothing other than one stage of disease that in due course of time will lead to serious developments. I tell you this with *double emphasis* because I believe that I know it to be the *truth*, and it could not be if the morning wants were not so easily disposed of. Nature will not permit any cutting down of her actual needs without a protest vocal with power. It is not a matter at all of getting her used to doing without a needed morning meal, for this she would never permit ; it is simply a means of getting over *one stage of disease* in the only way it can be done, by keeping the stomach always free from lunch-pail duties during the morning hours.

Another question.

DOCTOR,—What about the habitual drinking of large quantities of water simply for the purpose of "washing out" the stomach and the kidneys, and especially with waters that are supposed to be medicinal ?

(Signed)

MANY READERS.

In my own case I never take a drink of water except when athirst. My thirst is so rare that I am compelled to believe that Nature designed that I should get all the lime, soda, potash, magnesia, etc., that I need from the food I eat.

As to the washing-out process, as I understand it, every glass of water that goes into the stomach, in proportion to the need as indicated by thirst, becomes rapidly absorbed into the blood-vessels and hence *it* becomes diluted by the addition of the glass of water. There is never any sense of refreshment when water is taken without thirst. As water is simply absorbed there is not likely to be peristaltic motion excited unless in a very feeble way, hence the stomach is not likely to get much cleaner for the washing resulting, and as it all gets into the blood it cannot well reach the kidneys in "washing-out" quantities, and to any extent that this is possible, it must be at the cost of needless work of the kidneys themselves.

The marvelous cures alleged from cold-water means, are so involved with contributive agencies as to leave the question a matter of very crude inference. In this light I do not hesitate to strongly advise that you never take a drink of water except as Nature indicates, and then let it be as pure as available, and as cool as shall satisfy the "thirsty soul." Thirst never calls for tepid or hot water.

Now comes a very important question.

DOCTOR,—You have alluded several times to the use of apples, peaches, etc., in a way that seems to call for something more definite from you. After living all our lives under the impression that these fruits can be taken at any time of day with no harm because considered so "wholesome," here you come in with your doubts as to this cherished, ingrained idea!

Now you already believe that these fruits should not

be taken between meals, so we only need to discuss their use at meals.

Sir William Roberts in his elaborate work on "Digestion and Diet" tells us that there are potash salts in all green fruits in feeble combination with vegetable acids ; that when these fruits reach the stomach, the free muriatic acid which is one constituent of the gastric juice having a stronger affinity for the potash salt than the vegetable acid, at once combines and sets it free, and hence the gastric juice becomes weakened by the loss of this powerful solvent, and hence the sourness of the stomach that invites so much of soda to relieve it.

There is a very **large** per cent. of people in mature life who cannot eat **green** fruits without trouble, and I believe it to be due to this chemical change, but also that they are harder to digest than is commonly supposed.   I grant that they are a luxury of eating, but the pleasure of eating is only a short experience, while the pangs of indigestion are *long.*

In the case of my sons who do all of their eating at the table, even the finest exhibition of raw fruits has only the feeblest temptation.   What you shall eat when you get to the table with your rested stomachs is much more a matter of indifference than can ever be the case when no power has had a chance to accumulate.

The strong stomach, like the strong man, can recover from an over-tax with more rapidity and ease.

And here we may consider the question of the *bill of fare for the hungry.*

I am free to say to you that I know of nothing in the line of human duty so difficult as to avoid an excessive meal.   Hunger, with its keen sense of relish, is a most

potent tempter to excess, particularly as at every
" well-to-do " table there are those tempting desserts
to be considered when natural hunger has been fully
sated by food in excess of the limited supply of gastric
solvents.

In this there is need for every human being to call
on all within him in judgment and self-control, even
to the last reserve, that there shall not be eating far in
excess of the power to dispose of by the chemistry of
digestion.

There can be large liberty taken with the bill of fare
when the empty stomach is to be filled, but not the less
must sin be met where there is excess.

Then can be used these enticing persuaders to adding
a second meal before the first is arisen from, but the
consequences must inevitably be met.

One of you asks, What about the use of pork ? To
this I may say that it is one of the hardest, if not the
hardest, of meats to digest. And they are very few
who are engaged in mental labor who can use it without harm.

The manual laborer, he of the pick and shovel, can
deal with more power, because his life is so simple and
his business is so simple that there can be little difficulty in a shortage of nerve force for digestive processes while the labor is going on. His labor permits
rest between labor hours, and, as he has no worrying
over his affairs, he can both rest and sleep when not on
duty.

No man who has any fine brain work to do should
ever tax his stomach with ham or bacon on the same
day, as it is never a fit food for the brain worker.

One of you asks for still more light as to the dis-
tinction between morbid and natural hunger. Morbid

hunger is the morning hunger ; it is that hunger that is never satisfed with eating, that exists often between meals. *Morbid hunger is disease.*

Natural hunger is hunger in repose, that can wait longer easily if necessary—it is not attended with that nervous haste and impatience that incites bolting at the table. *Natural hunger is never in a hurry,* and hence it permits mastication and the highest flow of gastric juice, of the saliva, as the delicious experience of the table goes on.

LECTURE XXV.

My Friends the Ministers :—

I have invited a few of you to meet me in **special** session. I meet you with peculiar pleasure, not only because of my estimate of your moral, mental and personal worth, not only because your lives, your examples, your teachings make you in a peculiar sense the very salt of the earth, but because that from a child I was taught to revere and to honor the members of your profession. I was "born and bred" in a home where a minister was believed to be a "messenger." A man "called of God," who bore upon his person an actual commission, "signed, sealed, and delivered by the Divine hand."

In those older and more reverential times, the homes of the fathers in the church were in a peculiar sense the homes of the pastors, and of the traveling ministers ; and no guests were ever more welcome. They brought to the fireside a culture that was all the more marked because of its absence in the homes. The printing-press was not then making culture and knowledge available to all, and as they themselves were regenerated in a physical sense by the bread of life received, so did they freshen all the lives with whom they came in contact by administering bread to the moral and intellectual need.

There is no memory so sacred to me, not one so indelibly impressed as the memory of the expression that illumined the countenance of my parents when, by chance, the morning and evening devotions that were the abiding rule in the home were led by a "Man of God."

You, gentlemen, by virtue of your calling and worth, may well be considered the most striking illustrations of the possibilities of a civilization whose foundations rest upon the Book of books, the Bible. You are our examples, our teachers, our leaders in all that elevates and refines human character, and you should be able to so live, that your influence should be all-abounding, not only in this, but in the illustrative sense that comes from a more perfect physical manhood.

It is yours, gentlemen, to walk more by faith; to teach the law of God manifest in the soul; mine to walk more by sight, to teach the law of God manifest in the flesh. The lines of our lives are in close parallel, but do not run in the same groove. We are very near each other, each trying to do faithfully and conscientiously the work Nature designed each to do.

There can be no conflict between the science I am trying to unfold and to enforce, and the religion that you are trying to unfold, to teach, to enforce, except you yourselves invite it.

There is never any conflict incited by you, that does not come from ignorance of the law "according to the flesh" that is heathenish in its density.

You go into your pulpits with loaded stomachs to teach the word of life with a conflict invited between a chemical science and a religious teaching, between "*science and religion*," in which science is not often the worsted party.

And then we hear of your headaches, your exhaustion, and your biliousness, that seem to suggest a remedy. Ah, gentlemen, the liver, the stomach, the heart, the lungs, etc., all complain because the attacks have come from *without*, rarely because they themselves are primarily at fault. You, no less than your hearers, must suffer, because of your sins against the laws of the flesh. There is nothing for you in appeal to Divine aid that you may be supported in those trials through your painful afflictions, nothing in appeals for support, because your spiritual aim has fallen far below the mark, nothing more for you than for us, when we also suffer painfully and materially in our more practical affairs because of outraged law.

Nature, gentlemen, is a most exacting bookkeeper, and you cannot overdraw accounts, without resulting ceaseless painful reminders of reckless expenditures, and you suffer most who are habitually the most extravagant.

Now, as you have been among my listeners, it only remains for me to give you some hints as to the best way to so order your affairs as to be able to reach a far higher obedience to those laws of life, that are no less sacred, no less spiritual than the most elevated, the most refined that you are called upon to unfold to the world's need.

Let all of your purely intellectual work be done with your stomachs in absolute *natural repose*. Muffle your study-bell to all calls until the mind can be released from all tasks. This done in due time there will be a signal from the bookkeeper to attend to the adjustment of a lost balance. This attended to, then turn on the steam, the recreative force that *must, must* be if you are to become regenerated men.

For some hours engage yourselves only in that which will add the greatest available cheer of mind.

There was one minister of the gospel who was able to "break the bread of spiritual life" to his people effectually during thirty years, during which he broke the bread of physical life to his stomach but once during each twenty-four hours!

Do all of your study with stomachs empty, but when the time comes for a renewal of strength, I need not suggest to you that, by virtue of your calling, you above all men are tempted to exceeding sins at the table. There is always a vacant chair at the table for you at the homes of parishioners, no less than in the reception-room, and those tables never want the highest available reach in temptations to sin. Adam fell because he was tempted, and as he yielded, so do you, and so do you fall, and so are you driven from that Garden of your first estate that ought to be your sure heritage.

When that day comes every week, when there is the utmost mental tension, do not ever break your fast until a rest after your first service, and then let the break be exceedingly slight. The most feeble among you, if you could once be made to believe it possible, could make the taxing Sabbath a day of absolute fasting, and the praying and the preaching would strikingly gain in unction and ease. This statement is void of the least extravagance because, as I have so often insisted, the brain will be duly nourished, while the taxing work goes on, and will be functionally far stronger if there is no abnormal "conflict" in operation.

It would doubtless require some culture, some time to educate the minds of some of our ailing ones up to this idea, but the result would inevitably be reached on persistent effort.

I invite you all, gentlemen, to begin at once to eliminate the sins against the body. You ought to be our highest types of physical as well as of spiritual life. You cannot be our ideals in the latter, without the former. No man can be a good Christian without most *extraordinary and persistent and ceaseless effort who is not physically well.* Your headaches and all your ailings, no matter how many or where they are located, are mainly the results of lifelong avoidable culture of sins against the body and against the soul, which often have to be confessed "with groanings that cannot be uttered" as to duly interpret the wrong-doing within. I would have you believe that headaches and bilious attacks are never anything less than direct humiliating evidences of excesses *at the table.*

They are the voices of outraged nature that need not be heard, that never ought to be heard. Why, gentlemen, for more than ten years, by simply waiting every day upon hunger before eating, I have not failed to relish one substantial and one light meal every day. This means that I have had no headaches, no bilious attacks, not even a cold severe enough to interfere with this regularity, and all this with the taxing days and nights at the bedside of acute illness. You can have no conception, gentlemen, of the possibilities of the culture of health, of the possibilities of sustained health, until you have given this natural, this physiological method in life a most persistent trial. You should begin at once with the confidence that your success will be gauged by the thoroughness with which you habitually obey the divine laws of life.

But you are not to forget that when sin has been great and long-enduring so will time be required to expiate. Many of you are like farms that have been cropped for years without fertilizers, or like overdrawn bank-accounts where only the interest was intended for your use.   This you will have to take earnestly into account and therefore not expect miracles.

There can be promised only the results of persistent painstaking culture.   There will be no miracles by which you will escape the surgeon's knife, no healing of disease, naturally and inevitably fatal, for science is never absurd.

You are advised, you are admonished to begin at once to save your lives, not only for your own sake, but that your ministrations for the good of others may prove far more fruitful than otherwise can be possible.

LECTURE XXVI.

My Friends the Readers :—

We meet this morning to consider some questions raised by many of you, and also to bring this series of lectures to a close. I have been pleased with the closeness with which you have followed me through each one, and especially pleased that I have been called upon to consider questions you have raised with a view to draw out more fully whatever of light I might have as to points of seeming importance not duly elaborated.

Since my last lecture I have been consulted by a divine of great learning and eminence who has been growing in the grace of a better physical life for several months, and who is to grow on for some years to come, as to insomnia and bedtime hunger.

In the consideration of this question I shall reiterate as little as possible points already gone over, and state some things as physiological facts with something of a "thus-saith-the-Lord" emphasis, *rarely disbelieved.*

You have been thoroughly convinced, every one of you, that hunger will come if you patiently wait for it, even in any case of sickness where death is not inevitable, and as evidence of this, and to add still more emphasis to that which I have abundantly furnished you, I am fortunate in having a remarkable case for another illustration.

311

The city missionary of Norwich, Conn., has a son
in his fifth year whose appetite began to fail in 1893,
with a resulting languor and indisposition to be about
in the active sports of one of his years. He was duly
treated medicinally without avail. The symptoms so
developed that he entered the year 1894 with a stomach
that habitually ejected all meals some time after eating. This eccentric stomach was duly treated with
all the remedies apparently indicated, by both schools
of medical practice (homeopathic and regular), but
with not one hint of relief. There was no hunger,
but still the feeding went on with the highest science
of dietetics duly applied. The month of June came,
the boy was merging rapidly to the skeleton condition.
His parents had become resigned to the death that
seemed inevitable, but it is human in all cases to keep
on trying until the chill of death actually comes.

There remained one more effort; the boy was duly
enwrapped, and the father started with the wasted
form to an office where there was to be a general consultation of physicians. By a chance, a mere chance,
he met my friend, the publisher, who at once inquired
into the case.

Now it so happened that my friend had, as you have
been told, the experience of the case of Dr. Alexander,
who, reduced to ninety-seven pounds, was enjoined to go
upon a fast until hunger should come, no matter how
many days it might require. He had seen this man
actually gain strength on his fast of eight days, with a
marked decline of symptoms; he had seen an appetite
regained that came habitually at the same time every
day with the precision of a chronometer, and he was
bold enough, such was his faith, to advise that the boy
be taken back home and be permitted to rest, free from
all importunities, until Nature herself should ask, in

unmuffled tones, for what was needed to restore the waste due to the wreckage of enforced feeding.

The father was persuaded, the boy had four days of such rest as he had not known since he became ill, when, lo! there was a call, not for chemically-prepared food, but for *beefsteak!* It was given; it was duly digested, and the life was saved. From thence on there was no trouble.

This boy had suffered a living death week after week, month after month, and yet, at most, *always within four days* of a beefsteak appetite.

The following is the statement of the boy's father:

In June, 1894, my son had become reduced to a mere skeleton, and he suffered so much from soreness that changing his clothes was refrained from as much as possible, and it was necessary for him to rest on a pillow whenever he was taken out for a drive. All hope and apparently all reason for hope was gone, but there remained just one more thing to be done, a consultation of several physicians. But while tenderly bearing the wasted form along the crowded streets for this purpose, by the merest chance, the chance of a half a minute perhaps, the apparently absurd suggestion was met that the starved boy should be taken back home and be put upon a fast until hunger for substantial food should come. My supreme faith in the man who made the suggestion and the abundant external evidences in support of the faith within him was sufficient to turn me about—and my son was *raised from the dead.*

Signed,

Geo. W. Swan.

Dr. Alexander spent four years of his life in his home slowly wasting away with not one morsel of food

taken with relish, only praying that he might be so
blest as to die, that he might be relieved from the
pangs of a living death—and yet all the while, clearly,
safely, within nine days of a beefsteak appetite!

My friend, the publisher, spent eight years of his
life eating three meals daily when they could be enforced, hunting, hunting, ever hunting, for an appetite
that could not be found in a tireless search through
the old world, that could not be found among the hills
and mountains of his native country, and all the
while clearly, safely, easily within *three days* of such
an appetite as was never exceeded in the lustiest days
of his youth!

Ah, readers, how wonderful "Nature's bill of fare
for the sick," whereby it is absolutely safe, in any kind
of a case, to await Nature's demand for food until it
can be taken with such relish that eating becomes
a luxury!

In November, 1894, I had the pleasure of seeing the
little boy who, from his first relished meal, had been
able to eat from thence two substantial meals daily,
and he had all the weight and firmness of muscle, all
the health and cheer and activity that normal health
could make possible in his case.

I also had the pleasure of seeing Dr. Alexander, who
rode across the country twenty miles to tell me the
story of a life saved.   He had gained *nineteen pounds.*
Said he: "For twenty-five years I had been eating
with no appetite or a very poor appetite; since that
first hearty meal after the eight days' fast, I have not
failed of one keenly-relished meal every day," and the
ruddy skin, the bright eyes, the beaming expression
were more eloquent than any words he could utter.

For ten days I was the guest of my publisher, and
the results in his case and in his family seemed an

abundant reward for the years of posing in my native city as a conspicuous target for all the powers of language in ridicule, in epithet, in denouncement, while trying to unfold to my fellow-citizens those physiological laws involved in the culture and main-tenance of health, which, duly observed, was to make each and all happier, longer-lived, and therefore stronger physically, morally and mentally. But the day of ridicule is over.

From the first case of a life saved each has been a potent influence in the saving of other lives; the method has gone from one suffering friend to another suffering friend, from one family to another family, until at last I may walk my ways along the streets, as the instigator and a promoter of a revolution that will never go backward, because experience has proved, will prove, that it is based upon a larger compliance with those laws whereby we live, move and have our being.

Now I want to assure you who are the most thoughtful of my readers, whose lives are most gov-erned by reason, that bedtime hunger is never to be indulged. You will always take that second meal so long before you enter your beds that your stomachs shall have that same rest during the entire night that shall be granted all else that needs resting. And when for any reason the second meal is not duly handled, and there results bedtime hunger, you will always find an absence of it in the morning and will be glad you did not indulge it.

It is very true, however, that you will get to sleep sooner and sleep longer, if you do indulge the symp-tom; but there will be more of torpor than of restful sleep, you will always awaken with a feeling of ex-haustion such as occurs in after-dinner naps.

A stomach that can rest during the night is going to add a great deal to its power during the day. You are to consider insomnia as a disease, and to treat it by letting the stomach rest at night; eating, often, but not always, may give you more time of oblivion or of disturbing dreams, but never that refreshment, that regeneration that comes with sleep when the stomach is also in repose. It is the experience of all who have adopted this method in life, that sleep is so much more refreshing that not so many hours are required, and hence a decided gain in time, time invaluable for this one life upon earth, too short for all we would do, for all we would enjoy.

Our brightest thoughts are always on the brightest days; cloudy days and the darkness of night tend to gloom; "night thoughts" are generally gloomy thoughts, and by as much depressing to digestive energy. The imagination becomes dominant and controls the mind at will as the winds the ship that has lost its rudder. You can stand the loss of sleep with the stomach empty infinitely better than you can with nerve force engaged in a needless task of disposing of food without digesting it.

You who have the disease insomnia are admonished that it is to be cured by an evolution in reverse. You are vehemently advised to spend every minute when awake in your beds in thought over some subject vital to your lives, to your happiness. Get your minds under control so that you can do this.

During the many years which I have been engaged in thought over the subject matter of this course of lectures, it was my habit to spend every "waking hour" in bed in the elaboration of points, reserving the day for whatever of gloom that had to be met from avoidable or other cause. This became a fixed habit, and the result was no time lost through un-controlled imagination, and rarely less sleep than the normal demand, but certainly less sleep than the average during the many years of gluttonous feeding.

It occurs to me almost daily to defend this method in nature against arguments drawn from the habits of animals.

Said a learned judge, "The cow eats when hungry and seems to do well when following her natural instincts." Very true, judge, you will trust your favorite Jersey all day long to feed at will on the tender, watery grass of the pasture but not one hour in a field of clover ready for the scythe; not thirty minutes before an open grain bin. Animals will never fail to gorge themselves when opportunity occurs; it is only in time of sickness that their *instincts* rise superior to the *sense* of man.

Because animals are inclined to repose after meals, so it is believed that sleep should naturally follow eating. There can be no doubt that digestion is a tax on muscle energy in animals as well as in man. There is a digestive torpor; but who ever saw a cow or a horse asleep? I have even been reminded of the torpor of the anaconda during the days when his one huge mouthful is undergoing the digestive process, as an argument in favor of inviting sleep by eating, but does he actually sleep?

Can we compare the digestive processes of animals which are never influenced by mental conditions with those of man which are as sensitive to the conditions of the mind as the flame to the wind, as the leaf to the zephyr? *Absurd!*

The question is often asked me, whether it is safe for the old and infirm to abandon a lifelong eating habit for a method that requires that eating shall be regulated by the time of hunger, and not by the time of day. As well ask whether it is safe to abandon a lifelong habit of sin against a moral law by reason of age and infirmity. One can never get so old, never so infirm, as not to get an immediate change for the better, where there has been habitual sin against the Divine laws of digestion, and the older and the more infirm, so is the need more imperative.

It is often urged that by confining the stomach to the labor of digestion of only two meals daily, it will become unable to ever handle more meals during the same day. My dear readers, there is not one stomach in a thousand that can handle three meals a day and keep it up where there is any brain to interfere with digestive energy. You who eat, or try to eat, your three daily meals habitually, keep your stomachs weakened by over-work. You who eat only two meals daily, have the power to indulge the gluttony of a picnic dinner or of a church supper, your three meal friends know nothing about. The strong man armed has the greatest power in defense.

And now it remains for me to admonish you that new lives are before you if you get yourselves in line with Nature's laws ; lives that may be freed from the taxing, the perplexing cares as to what you are to do or what not to do that you may be healthy. You are to rise every morning with your rested brains, muscles and stomachs and go about your affairs with high hopes, with confident expectations that while you await the development of hunger, there can be no development of disease going on at the same time, where you are not already victims under the aim of the drawn bow. You are now aware that disease is no less a matter of culture than of health.

As time goes on with you, you will see more and more clearly, that your ailings are directly traceable to some avoidable error that will tend to more and more care as to the violation of law. It will be an immense gain to your lives to be freed from your superstitious fears of disease, and your superstitious needs that they be treated, that they be exorcised by the oblations, the sacrifices to be burned upon an altar erected to an "unknown God." That altar shall not be the hapless stomach as you become more and more enlightened through experience, through a clearer apprehension of the means by which you live, move and may possess more vigorous, disease-defying bodies.

You shall walk your ways in this world drawing the breath of life when and where you will, regardless of the millions of microbes which are drawn in only to be expelled, because there is no lodging-place for them in tissues, *permeated, vivified, electrified* with that rich warm blood that comes from the vigorous digestion of food by stomachs that have regained their lost powers.

And you who are old and have suffered long from many infirmities, whose stomachs have been the altars for the burning of sacrifices innumerable and fruitless, there is hope for even you. If the "unknown God" has power to heal your wounds and broken bones, so has He power, no matter how old, how infirm you are, to add growth to your waste places.

Go then to those happier ways in life that shall make you better citizens, better parents, better children. Obey the Divine laws of your being and your days, your months, your years shall be long in the land the Lord thy God giveth thee.

LECTURE XXVII.

"Truth, crushed to earth, will rise again;
The eternal years of God are hers;
But Error, wounded, writhes with pain,
And dies among his worshipers."

Brothers in the Healing Art :—

The following quotation from Oliver Wendell Holmes was read to my class by our professor of theory and practice as an extravagance of a literary mind :

"Throw out opium, which the Creator Himself seems to prescribe, for we see the scarlet poppy growing in the cornfields, as if it were seen that whenever there is hunger to be fed there must also be pain to be soothed ; throw out a few specifics which our art did not discover and the vapors which produce the miracle of Anæsthesia and I firmly believe that if the whole materia medica, *as now used,* could be sunk to the bottom of the sea, it would be all the better for mankind—and all the worse for the fishes."

This was the only vocal utterance from the professor that was worth my while to remember, and it at once became a keynote, which, in the earlier years of my practice, when gravely in doubt as to whether any medicine were needed, broke in upon the "rapt porch of my ear," not as a "doleful sound," but as a voice of nature, soft, melodious, subduing, uplifting ; and it has lost none of its melody after thirty years of listening.

320

This heresy was uttered to the startled ears of the Massachusetts Medical Society in 1860 by Holmes the philosopher, the wit, the physician, the professor of anatomy, and yet his resignation was not called for. He further says :—

"But to justify this proposition, I must add that the injuries inflicted by over-medication are often masked by disease."

Dr. Hooker believes that the typhus syncopalis of preceding generations in New England was often "in fact a brandy and opium disease." How is a physician to distinguish the irritation of a blister from that caused by the inflammation it was meant to cure ? How can he tell the exhaustion produced by his evacuants from the collapse belonging to the disease they were meant to cure ? Brothers, these morning lectures are to be spread broadcast over the world by a consummate master in all the arts of advertising (the publisher), who is also a man of tireless energy and of the largest faith that they contain truths that ought to be known by all peoples. That they are to meet criticism, ridicule, remorseless, relentless, I know full well, for such has been their history all through the years of unfolding and giving to the people, until the author has become "case-hardened."

But the people will accept the principles and the practices enunciated when they understand them, and once a lodgment is gained in any cross-roads, village or city a "revolution" will begin that will never "go backward," and hostile criticism will not avail. Says Holmes in his preface, "The character of the opposition that some of these papers have met suggests the inference that they contain really important but unwelcome truths. Hostile criticisms meeting together are often equivalent to praise, and the square of fault-finding turns out to be the same thing as eulogy."

21

As the poet, the philosopher, the scholar, the physician, gave me a keynote by which a course of lectures was made possible, so shall he end them.

" MY BROTHERS IN THE ART :

" There is nothing to be feared from the utterance of any seeming heresy to which you have listened. I cannot compromise your collective wisdom. If I have strained the truth one hair's breadth for the sake of an epigram or an antithesis, you are accustomed to count the normal pulsebeats of sound judgment, and know full well how to recognize the fever throbs and nervous palpitations of rhetoric.

"The freedom with which each of us speaks his thought in this presence belongs in part to the assured position of the profession in our commonwealth, to the attitude of science, which is always fearless, and to the genius of the soil on which we stand, from which Nature withheld the fatal gift of malaria only to fill it with exhalations that breed the fever of inquiry in our blood and in our brain. But mainly we owe the large license of speech we enjoy to those influences and privileges common to us all as self-governing Americans.

" This Republic is the chosen home of *minorities*, the greater material powers have always ruled before. The history of most countries has been that of majorities—mounted majorities clad in iron, armed with death, treading down the tenfold more numerous minorities. In the old civilizations they root themselves like oaks in the soil : men must live in their shadow or cut them down. With us the majority is only the flower of the passing noon, and the minority is the bud which may open in the next morning's sun. We must be tolerant, for the thought which stammers on a single tongue to-day may organize itself in the growing consciousness of the times, and come back to us like the voice of the multitudinous waves of the ocean on the morrow."